Lecture Notes in Computer Science   13172

More information about this subseries at https://link.springer.com/bookseries/7410

Weizhi Meng · Mauro Conti (Eds.)

Cyberspace Safety and Security

13th International Symposium, CSS 2021
Virtual Event, November 9–11, 2021
Proceedings

 Springer

Editors
Weizhi Meng ⓘ
Technical University of Denmark
Kongens Lyngby, Denmark

Mauro Conti ⓘ
University of Padua
Padua, Italy

ISSN 0302-9743   ISSN 1611-3349 (electronic)
Lecture Notes in Computer Science
ISBN 978-3-030-94028-7   ISBN 978-3-030-94029-4 (eBook)
https://doi.org/10.1007/978-3-030-94029-4

LNCS Sublibrary: SL4 – Security and Cryptology

This Springer imprint is published by the registered company Springer Nature Switzerland AG
The registered company address is: Gewerbestrasse 11, 6330 Cham, Switzerland

Preface

This book contains the papers that were selected for presentation and publication at the 13th International Symposium on Cyberspace Safety and Security (CSS 2021), which was organized by the Cyber Security Section of the Technical University of Denmark, Denmark, held during November 9–11, 2021. CSS 2021 was the latest edition in a series of highly successful international events on cyberspace safety and security previously held as CSS 2020 (Hainan, China), CSS 2019 (Guangzhou, China), CSS 2018 (Amalfi, Italy), CSS 2017 (Xi'an, China), CSS 2016 (Granada, Spain), CSS 2015 (New York, USA), CSS 2014 (Paris, France), CSS 2013 (Zhangjiajie, China), CSS 2012 (Melbourne, Australia), CSS 2012 (Milan, Italy), CSS 2009 (Chengdu, China), and CSS 2008 (Sydney, Australia). Due to COVID-19, CSS 2021 was held online.

This year's Program Committee (PC) consisted of 42 members with diverse background and broad research interests. A total of 35 papers were submitted to the conference. Each paper was evaluated on the basis of its significance, novelty, and technical quality. Most papers were reviewed by three or more PC members. Finally, nine full papers were selected for presentation at the conference with an acceptance rate of 25.7%. In addition, five short papers were accepted.

CSS 2021 had six outstanding keynote talks: "Em-Curve25519: Faster and Smaller" presented by Zhe Liu from the Nanjing University of Aeronautics and Astronautics, China, "Achieving Efficient and Privacy-Preserving Dynamic Skyline Query in Online Medical Diagnosis" presented by Rongxing Lu from the University of New Brunswick, Canada, "Artificial Intelligence for Biometric Technologies and Systems" presented by Vincenzo Piuri from the Universita degli Studi di Milano, Italy, "The Quest for Industrial Host Security" presented by Nils Ole Tippenhauer from the CISPA Helmholtz Center for Information Security, Germany, "Cyber-Physical-Social Systems: Design, Analytics, Security and Privacy" presented by Laurence T. Yang from Hainan University, China/St. Francis Xavier University, Canada, and "What do I Trust? Securing Critical Infrastructure in Cyber War" presented by Jianying Zhou from the Singapore University of Technology and Design, Singapore. Our deepest gratitude for their excellent presentations.

For the success of CSS 2021, we would like to first thank the authors of all submissions and all the PC members for their great efforts in selecting the papers. We also thank all the external reviewers for assisting the reviewing process. For the conference organization, we would like to thank the CSS Steering Committee chair – Yang Xiang, the general chairs - Nicola Dragoni and Jun Zhang, the publicity chairs - Alessandro Brighente and Wenjuan Li, and the publication chair – Wei-Yang Chiu. Finally, we thank everyone else, speakers and session chairs, for their contribution to the program of CSS 2021.

November 2021

Weizhi Meng
Mauro Conti

Organization

General Chairs

Nicola Dragoni   Technical University of Denmark, Denmark
Jun Zhang   Swinburne University of Technology, Australia

Program Chairs

Weizhi Meng   Technical University of Denmark, Denmark
Mauro Conti   University of Padua, Italy

Publicity Chairs

Alessandro Brighente   University of Padua, Italy
Wenjuan Li   Hong Kong Polytechnic University, China

Publication Chair

Wei-Yang Chiu   Technical University of Denmark, Denmark

Special Issue Chair

Yu Wang   Guangzhou University, China

Steering Committee Chair

Yang Xiang   Swinburne University of Technology, Australia

Technical Program Committee

Andrea Abate   University of Salerno, Italy
Cristina Alcaraz   University of Malaga, Spain
Silvio Barra   University of Naples, Italy
Gergely Biczok   Budapest University of Technology and
     Economics, Hungary
Chao Chen   James Cook University, Australia
Ashutosh Dhar Dwivedi   Technical University of Denmark, Denmark
Changyu Dong   Newcastle University, UK
Sven Dietrich   City University of New York, USA

Dieter Gollmann	Hamburg University of Technology, Germany
Chunpeng Ge	Nanjing University of Aeronautics and Astronautics, China
Julian Jang-Jaccard	Massey University, New Zealand
Haibo Hu	Hong Kong Polytechnic University, China
Jinguang Han	Queen's University Belfast, UK
Qiong Huang	South China Agricultural University, China
Costas Lambrinoudakis	University of Piraeus, Greece
Wenjuan Li	Hong Kong Polytechnic University, China
Jiqiang Lu	Beihang University, China
Shigang Liu	Swinburne University of Technology, Australia
Jay Ligatti	University of South Florida, USA
Lorena Gonzalez Manzano	Universidad Carlos III de Madrid, Spain
Vincenzo Moscato	University of Naples, Italy
David Naccache	École normale supérieure (ENS), France
Lei Pan	Deakin University, Australia
Claudia Peersman	University of Bristol, UK
Josef Pieprzyk	CSIRO/Data61, Australia
Indrajit Ray	Colorado State University, USA
Jun Shao	Zhejiang Gongshang University, China
Chunhua Su	University of Aizu, Japan
Zhiyuan Tan	Edinburgh Napier University, UK
Zhe Xia	Wuhan University of Technology, China
Peng Xu	Huazhong University of Science and Technology, China
Lei Xue	Hong Kong Polytechnic University, China
Wun-She Yap	Universiti Tunku Abdul Rahman, Malaysia
Tsz Hon Yuen	University of Hong Kong, China
Yong Yu	Shaanxi Normal University, China
Ding Wang	Nankai University, China
Qianghong Wu	Beihang University, China
Xuyun Zhang	Macquarie University, Australia
Tianqing Zhu	University of Technology Sydney, Australia

Additional Reviewers

Shangbin Han
Shuaishuai Liu
Meiyan Xiao

Contents

Encrypted Malicious Traffic Detection Based on Ensemble Learning

Fengrui Xiao[1], Feng Yang[1,2]($\boxtimes$), Shuangwu Chen[1,2], and Jian Yang[1,2]

[1] Department of Automation, University of Science and Technology of China,
Hefei, China
franxfr@mail.ustc.edu.cn, {yangf,chensw,jianyang}@ustc.edu.cn
[2] Institute of Artificial Intelligence, Hefei Comprehensive National Science Center,
Hefei, China

Abstract. Nowadays, network traffic detection plays a very important role in protecting cyberspace security, and more and more applications realize data privacy protection through encryption technology. Regular expression matching based methods, such as deep packet inspection that relies on plaintext traffic cannot be applied to detecting encrypted random communication content, and the existing detecting methods based on time-series features often ignore the encryption protocol features. In this work, we design an ensemble learning system based on stack algorithms to identify encrypted malicious traffic, which can detect the interactive behavior and the encryption protocols simultaneously. In detail, we construct a deep learning classifier based on Long Short-Term Memory (LSTM) for time-series features, and a machine learning classifier based on random forests for encryption protocol features. Then, we use the stacking algorithm in ensemble learning to combine them to form a new classifier. Finally, relying on the Datacon2020 dataset, extensive experiments are conducted. The experimental results indicate that the proposed method improves the detection rate of encrypted malicious traffic while keeping a low false positive rate.

Keywords: Malicious traffic detection · Encrypted traffic · Network security · Ensemble learning

1 Introduction

With the increasing awareness of risk prevention on the Internet, more and more applications realize data privacy protection through encryption technology, and the proportion of encrypted traffic in the network is growing continuously [20]. According to Google's statistics [18], the proportion of encrypted pages loaded by Chrome has reached 95% in 2019. Despite these needs and benefits,

This work was supported by the National Key R&D Program of China (No. 2018YFF01012200), and the Key Science and Technology Project of Anhui (No. 202103a05020006), and the Anhui Provincial Natural Science Foundation (No. 1908085QF266).

the encryption of traffic poses additional challenges to the detection of malware. Attackers can hide their malicious behavior through encryption, such as Trojan viruses, CC attacks, and malicious applications. However, the traditional expression matching based methods, such as deep packet inspection, which relies on plaintext traffic [14,21] cannot deal with the encrypted random communication content [24].

To address the above issues, recent researches mostly concentrate on developing machine learning or deep learning based encrypted malicious traffic detection methods, which rely on the statistical and interaction features that are not affected by encryption. However, these methods have the following limitations: 1) Since a network flow may last for a long time, the statistical features based methods typically focus only on the first few packets during the session estimation (e.g. SSL handshake) [11,12,15,26], without consideration of the behavioral features during the following data transmission. An attacker can easily bypass such detection methods by pretending to be normal at the session estimation stage. 2) The method based on TLS/SSL protocol features only captures coarse-grained features [7,22] but ignores the content of the application data. For example, when the encryption protocols and certificates used by two different flows are close, their protocol features will be considered the same. Therefore, the two flows are likely to be misclassified into one category, but the message information they transmit may be completely different.

In this paper, we design an ensemble learning framework based on stacking algorithm to identify encrypted malicious traffic. This method combines sequence-based LSTM model with traditional machine learning method, which can not only describe the message interaction, but also represent encryption protocol attributes. The LSTM model also uses the attention mechanism to enhance its ability to process long sequences. And the addition of ensemble learning algorithms allows our proposed method to deeply characterize the behavior of attackers and achieve efficient identification of malicious traffic.

The contributions of this paper can be briefly summarized as follows:

- We design an encrypted malicious traffic detection system, which takes both the length-related feature and the encryption protocols features into consideration.
- We propose an ensemble learning-based malicious traffic detecting method, which can detect the interactive behavior and the encryption protocols simultaneously.
- We carry out the experiments on real-world network traffic data for the encrypted traffic classification. The results show that our method effectively improves the detection rate of encrypted malicious traffic while keeping a low false positive rate.

The rest of the paper is organized as follows. Section 2 summarizes the related work and the detailed system architecture is proposed in Sect. 3. Section 4 presents the comparison experiments. Finally, we conclude this paper in Sect. 5.

2   Related Work

The detection of encrypted malicious traffic has become an important issue, which has received extensive research and attention in the field of industry and academia. The existing work can generally be divided into two categories: conventional machine learning methods and deep learning methods.

2.1   Malicious Traffic Detection Based on Machine Learning

The latest research written by Cao et al. [6] introduced the advantages and disadvantages of commonly used methods. In [1], Alshammari et al. studied two commonly used encryption traffic identification methods, namely expert-driven system and data-driven system, and concluded that the data-driven system is superior to the expert-driven system in terms of detection accuracy and false alarm rate. In [22], Taylor et al. proposed AppScanner that uses support vector classifiers and random forest classifiers to automatically identify mobile applications. By observing unencrypted TLS handshake messages and using several common machine learning methods, [2] provides a comprehensive study of the use of malicious TLS. Combining the Markov chain with the hidden Markov model, a weighted ensemble strategy is designed in [16]. Liu et al. [12] introduce the concept of Length Block sequence, and propose a method named multi-attribute Markov probability fingerprints (MaMPF).

2.2   Malicious Traffic Detection Based on Deep Learning

Lotfollahi et al. [13] used a stacked autoencoder and a one-dimensional convolutional neural network to automatically extract features from the encrypted traffic payload. Rui et al. [10] provided a byte segment neural network for traffic classification, put the payload segment into the attention encoder to obtain features, and then use the softmax classifier to classify. Chang Liu et al. [11] designed an end-to-end encrypted flow classification model called FS-Net. Haipeng Yao et al. [26] uses attention-based LSTM and hierarchical attention network (HAN). By combining LSTM autoencoder with deep dictionary learning, Junchi Xing [25] proposed a semi-supervised anomaly detection method. In [19], Meng Shen et al. present GraphDApp, a novel DApp fingerprinting method using Graph Neural Networks (GNNs), to implicitly reserves multiple dimensional features in bidirectional client-server interactions.

In summary, existing encryption malicious traffic detection methods, including machine learning methods and deep learning methods, input the extracted features as a whole into their system, but have not noticed the difference in these feature categories. In our work, we used different methods to process different features and combined the results for final identification.

3   Encrypted Malicious Traffic Detection Based on Ensemble Learning

3.1   Overview of Proposed Method

Although most of the network traffic is encrypted, we can still detect malicious traffic through some features that can reflect the behavior of the message senders. These features can be roughly divided into: statistical features and protocol features.

Statistical features refer to the sequence of data packet length or time interval. Generally, malicious communication data flows have the characteristics of infrequent interaction and more uploaded data [2]. Based on this difference in interaction behavior, although the packet length and arrival time cannot provide information about the content of the session, they can still facilitate inferences about sessional behavior. Since such features are all sequences and LSTM model has a good performance in processing this kind of data, we design an LSTM classifier with attention mechanism to deal with this type of features. Protocol features refer to the features extracted from the TLS/SSL encryption protocol metadata, such as the number of cipher suites provided, etc. These features are helpful to understand the behavioral preferences of the entire encrypted session, but not suitable for using sequence-based method to deal with. We use random forest method for processing this kind of features.

An overview of the method framework proposed in this paper is shown in Fig. 1, which is mainly composed of three parts.

- **Feature Extractor**: This module preprocesses the raw data input, i.e. extracts the features of the flows. The extracted features include: the statistical features of the flows and the encryption protocol features of the flows, which contains the TLS/SSL message information.
- **Primary Classifiers**: The stacking algorithm contains two-layer classifiers. This module is the first layer classifiers of the ensemble learning model and composed of two classifiers, which are the LSTM classifier with attention mechanism to deal with statistical features and the random forest classifier to deal with encryption protocol features.
- **Secondary Classifier**: This module is the second layer classifier of the stacking algorithm. In our work, we used a logistic regression model to process the metadata generated by the primary classifiers.

3.2   Feature Extracting

There are many ways for malware to perform encrypted communication. This article takes the most common encrypted communication based on TLS/SSL as an example. As mentioned before, there are two kinds of traffic features that can be used to identify encrypted traffic. For protocol features, we extracted 19 handshake protocol features from the encryption protocol header of the packet, including the number of Encrypted Alert packets, the number of SSL Client

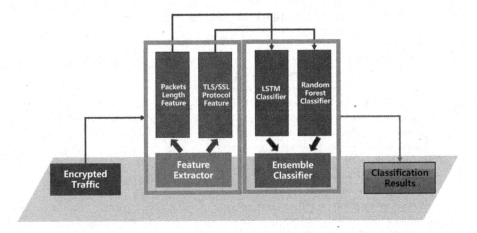

Fig. 1. An overview of encrypted malicious traffic detection system

Hello packets, etc. For statistical features, we mainly consider the length-related features of the flow, including four parts: (1) Uplink packets length sequence; (2) Downlink packets length sequence; (3) Uplink application layer packets length sequence; (4) Downlink application layer packets length sequence. These features are related to the state of the network between communication nodes and the running of the software. They are not affected by encryption protocols and are fairly stable even when the malicious is updated. The complete feature list is given in Table 1.

3.3   Primary Classifier for Statistical Features

Since the length-related statistical features of network traffic are time series data, this section uses an improved RNN model for modeling. Because the number of packets in the raw traffic data is very large, the sequence length we get is relatively long, and the traditional RNN model often faces the problem of gradient vanishing and gradient explosion when processing long sequence data, so we adopt the LSTM model [9]. LSTM can delete or add information to the hidden state vector with the help of a gate function. This means that LSTM can retain important information in the hidden layer vector. For $seq_i = \{x_1, x_2, ..., x_m\}, i = 1, 2, 3, 4$, the hidden vector can be obtained by following equation:

$$h_t = \kappa_\phi \left(\vec{x}_t, c_{t-1} \right) \tag{1}$$

where κ_ϕ is the LSTM kernel function and c_{t-1} is the memory cell at step $t-1$.

However, due to the large length of data flow, the LSTM model cannot remember all the information. In order to solve the long-term dependence of time series data, we use an attention mechanism [3] for the seq2seq model to calculate the weights of all hidden vectors, and finally get a weighted hidden vector, as shown below:

$$u_t = \tanh \left(W_p h_t \right), \tag{2}$$

Table 1. Feature list

Description	Category
Uplink packets length sequence	Statistical feature
Downlink packets length sequence	
Uplink application layer packets length sequence	
Downlink application layer packets length sequence	
Number of encrypted alert packets	Protocol feature
Number of SSL client hello packets	
Number of cipher suites provided by the client	
Maximum life cycle of non-CA certificate	
Number of client hello extension	
Server hello extension quantity	
Whether the certificate is self-signed or not	
Maximum lifetime of the certificate	
Maximum certificate chain length	
The name of the first server certificate owner	
The serial number of the 1st cipher suite provided by the client	
The serial number of the 2nd cipher suite provided by the client	
The serial number of the 3rd cipher suite provided by the client	
The serial number of the 1st cipher suite provided by the server	
The serial number of the 2nd cipher suite provided by the server	
The serial number of the 3rd cipher suite provided by the server	
Length of the 1st certificate	
Length of the 2nd certificate	
Length of the 3rd certificate	

$$\alpha_t = \frac{\exp\left(u_t^T u_s\right)}{\Sigma_j \exp\left(u_j^T u_s\right)}, \tag{3}$$

$$c = \sum_t \alpha_t h_t \tag{4}$$

where W_p, b_p, and u_s are the parameters that need to be trained, u_t is the importance score of each data packet, α_t is the normalized weight and $\sum_t \alpha = 1$.

The structure of the classifier based on LSTM model with attention mechanism is shown in the Fig. 2. The input of the classifier is the length-related statistical features mentioned in the previous section. Since we extract four sequential features, we use four LSTM units to vectorize each sequence into a hidden vector h_i, $i = 1, 2, 3, 4$. Each hidden vector is weighted by the attention model into a vector c_i, $i = 1, 2, 3, 4$. Finally, the softmax layer will output the classification result, while taking c_i as the input of the MLP network. The pseudo-code of Algorithm 1 is shown in Algorithm 1.

3.4   Primary Classifier for Protocol Features

Most user-level applications use popular TLS libraries for security purposes, such as BoringSSL (Chrome), NSS (Firefox) or SChannel (Internet Explorer).

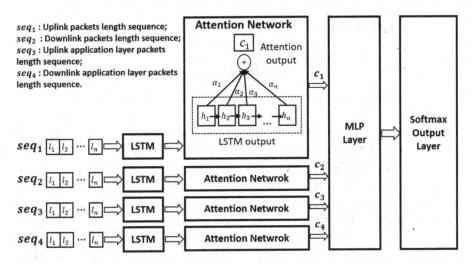

Fig. 2. The structure of the classifier based on LSTM model with attention mechanism

---

Algorithm 1. LSTM Model with Attention Mechanism

Input: The packets length sequences of network flow $S_{flow} = \{seq_i | i = 1, 2, 3, 4\}$;
Output: The detecting results of S_{flow};
1: **for** seq_i in S_{flow} **do**
2:    $h_i = LSTM_i(seq_i)$
3:    $c_i = attention\_network(h_i)$ (Eq.7-Eq.9)
4: **end for**
5: $x = [c_1, c_2, c_3, c_4]$
6: $result = softmax(MLP(x))$
7: **return** $result$

These applications usually have unique TLS fingerprints because developers will modify the default values of the library to optimize their applications. Thus, the TLS/SSL protocol header provides some very useful information, which can be used to distinguish different TLS libraries and applications. In addition to fingerprint features, the client usually will provide the server with a list of suitable cipher suites sorted by the client's preference. And each cipher suite defines a set of methods required to establish a connection and transmit data using TLS, such as encryption algorithms and pseudo-random functions. The client can also publish a set of TLS extensions, which can provide the server with the parameters required for key exchange, e.g. ec_point_formats [2].

Essentially, these features are the attributes of the network flows, and the decision tree algorithm has a good effect and interpretation for the classification of such features. Random forest is an ensemble learning algorithm that integrates multiple decision trees [5]. Based on the decision tree as the foundational learner, random forest further introduces random attribute selection in the training process of the decision tree. The integrated output of random forest is the average

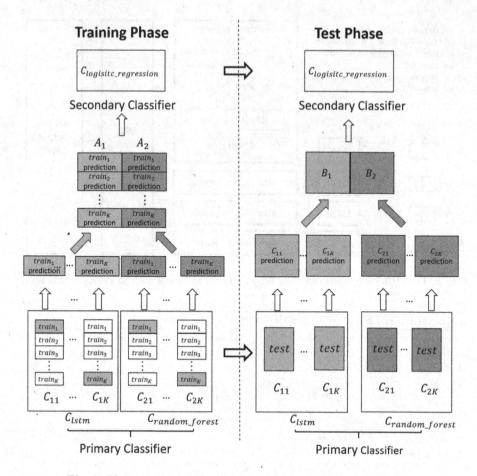

Fig. 3. The structure of the classifier based on stacking algorithm

probability of all individual decision trees. Although this method increases the deviation component of a single tree, the mean square error of the output is significantly reduced, which usually obtains superior performance.

3.5   Ensemble Classifier Based on Stacking Algorithm

The previous two sections respectively introduced the classification methods corresponding to their features. It's significant to combine the two classifiers. When the amount of training data is large, a very powerful combination strategy is to use the "learning method", i.e. to combine through another classifier. Stacking algorithm is a typical representative of "learning method" [4]. The original independent classifiers are called primary classifiers, and the classifiers used for combining are called secondary classifiers or meta-classifiers. Compared with ensemble strategies such as random forest and boosting, the primary classifiers

Algorithm 2. Framework of Ensemble Learning for Our System.

Input:
    The set of network flow $D = \{D_{train}, D_{test}\}$;
    Primary classifiers list:$C=[C_{lstm}, C_{rf}]$;
    Secondary classifiers:C_{lr};
    Cross-validation parameters:K;
Output:
    The detecting results of D;
1: $Feature = \varnothing$
2: **for** $Flow$ in D **do**
3:    $Feature \cup FeatureExtractor(Flow)$
4: **end for**
5: $Feature_{train} \rightarrow train_1, train_2, ..., train_K$
6: **for** i in range($len(C)$) **do**
7:    **for** j in range(K) **do**
8:       $C_{ij} = C_{ij}.train(Feature_{train} \setminus train_j)$
9:       $A_{ij} = C_{ij}.predict(train_j)$
10:     $B_{ij} = C_{ij}.predict(Feature_{test})$
11:    **end for**
12:   $A_i = A_{i1} \cup A_{i2} \cup ... \cup A_{iK}$
13:   $B_i = mean(B_{i1}, ..., B_{iK})$
14: **end for**
15: $A = (A_1, A_2)$
16: $B = (B_1, B_2)$
17: $C_{lr} = C_{lr}.train(A)$
18: $result = C_{lr}.predict(B)$
19: **return** $result$

of the stacking algorithm can combine different algorithms, which fits our situation.

According to Stacking algorithm, we first train the primary classifiers using the initial dataset, and then generates a new training set, which is usually called metadata, for training the secondary classifier. The output of the primary classifier is used as the instance input feature, and the label of the initial instance is still used as instance label. In the training phase, since the secondary training set is generated by the primary classifiers, the risk of overfitting is quite high. K-fold cross-validation is generally used to solve this problem, i.e. for each initial classifier, we divide the initial training set into K parts, and then train K classifiers. Researches have shown that using Logistic Regression as secondary classifier has better performance [23].

The overall procedures are described as follows, as shown in Fig. 3 and the pseudo-code of Algorithm 2 is shown in Algorithm 2. The training set is divided into K parts at first, denoted by $train_1, train_2, ..., train_K$. For the primary classifiers, take the neural network classifier as an example, use $train_1$, $train_2, ..., train_K$ in turn as the validation set, and the remaining $K - 1$ copies are used as the training set for K-fold cross-validation for model training.

And then make predictions on the test set, as shown in line 8–10. In this way, we get the K predictions on the respective validation set and the prediction B_{1j} on the test set. Then the K predictions are merged into A_1 and the average of sequence $(B_{11}, ..., B_{1K})$ is calculated as B_1. The random forest classifier does the same operation. After the two primary classifiers are trained, the predicted values of the two models on the training set are used as two "features", A_1 and A_2 respectively, for the training and modeling of the secondary classifiers. We use the trained secondary classifier model to make predictions on the values (B_1, B_2) which are constructed by the predicted values of the primary classifiers on the test set to obtain the final predicted results.

4   Evaluation

4.1   Experimental Environment

We use the pytorch library as the deep learning software framework [17], and the sklearn library to implement machine learning method. Our experiments run on CentOS 7.9 64 bits OS. The server CPU model is Intel(R) Xeon(R) Gold 5120@3.40 GHz, 64 GB of RAM. A Nvidia Telsa P4 GPU with 8 GB memory is used as an accelerator.

4.2   Dataset

Table 2. The structure of dataset

Category	Training set	Testing set	Overall
Malicious samples	3000	2000	5000
Benign samples	3000	2000	5000
Overall	6000	4000	10000

In the experiment, we used the dataset provided by the 2020 Datacon Security Competition [8], which is derived from malicious and normal software collected from February to June 2020. The malicious traffic in this dataset is encrypted traffic generated by malicious software, and white traffic is encrypted traffic generated by normal software. The traffic content is TLS/SSL data packets generated on port 443. Each sample in the dataset represents a flow receiving client, and the sample data contains flow data communicated between this client and one or more servers. The dataset structure is shown in Table 2.

4.3   Experiment Evaluation

Parameters Setting. The parameters of our experiment are set as shown in Table 3. Specifically, since FS-net uses a bi-GRU, the sequence length is 64 (32 * 2).

Table 3. Parameter setting of experiments

Model description	LSTM model	FS-net	Random forest	Xgboost	Proposed method
Parameters	num of layers: 4 hidden size: 64 time step: 32 learning rate: 0.0005 batch_size: 200	num of layers: 45 hidden size: 48 time step: 32 learning rate: 0.0005 batch_size: 200	n_estimator: 2000 max_leaf_nodes: 10	gamma: 0.1 max_depth: 6 min_child weights: 3	K-fold: 10

Table 4. Experimental results on precision, recall and F1-score

Method	Experiment results		
	Precision	Recall	F1-score
Decision tree	0.8565	0.4687	0.6059
Xgboost	0.7065	0.8074	0.7536
Random forest	0.8842	0.8400	0.8615
Support vector classification	0.8834	0.7845	0.8310
FS-net	0.8681	0.7845	0.8310
LSTM without attention	0.8408	0.8635	0.8520
LSTM with attention	0.9273	0.8480	0.8859
Proposed method	**0.9505**	**0.8635**	**0.9049**

Evaluation Metrics. Given the raw encrypted traffic, our goal is to identify whether it is malicious communication traffic. We have adopted three commonly used metrics for anomaly detection. Namely, *Precision*, *Recall*, and F_1*_score*, which are defined as follows:

- $Precision = \frac{TP(TruePositives)}{TP+FP(FalsePositives)}$

- $Recall = \frac{TP}{TP+FN(FalseNegatives)}$

- F_1_score $= \frac{2 \times Precision \times Recall}{Precision + Recall}$

Benchmarks. We compare our ensemble learning method with several other common methods. These methods include: 1) machine learning methods only using TLS/SSL protocol features, such as decision trees, SVM, xgboost, and random forest; 2) deep learning methods only using statistical features, i.e. FS-net [11] and LSTM models [26].

4.4   Experimental Results

The comparison results are shown in Table 4. From the results, we can see our method reaches high accuracy ($F_1 score \approx 90.5\%$) and outperforms other comparison methods on every metric. The two classifiers integrated by our method (Random Forest and LSTM model with attention mechanism) achieved the best

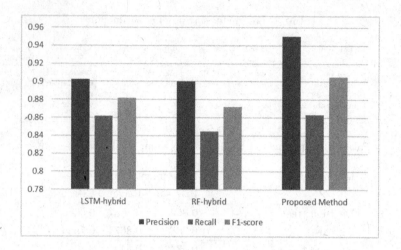

Fig. 4. Effects of stacking ensemble learning

results among the compared methods (excluding our proposed method). The Stacking algorithm requires that the primary classifiers are preferably strong learners and have good classification effects, otherwise the barrel effect will occur. And because the two classifiers use non-overlapping features, the two classifiers are independent of each other. Therefore, the classifier integrated by the two models has achieved better results. And the result shows that the performance of deep learning methods is slightly better than machine learning methods. This is probably because statistical features have more dimensions than protocol features and can better reflect users' data behavior. Moreover, deep learning algorithms are more complex and more efficient than machine learning. The attention mechanism also shows its effectiveness, increasing the $F_1 score$ of the LSTM model from 0.85 to 0.88.

In order to prove the effectiveness of ensemble learning with different classifiers, we simply concatenate all the features and input them into the respective primary classifiers for comparison. We implemented the following two algorithms as benchmarks: 1) Splicing length-related sequences with protocol features and inputting them to the classifier based on LSTM model with attention mechanism, which we call LSTM-hybrid; 2) Splicing length-related sequences with protocol features and inputting them to the random forest classifier, which we call RF-hybrid. The experimental results are shown in Fig. 4.

The results show that the ensemble learning classifier outperforms the other methods. Compared to the LSTM model with attention mechanism only using statistical features, the effect of LSTM-hybrid even decreased slightly. The results indicate that the difference between the two types of features has a negative impact on the neural network model. Through stacking integrated learning, the advantages of the two types of algorithms are well absorbed, and hence produces a good improvement.

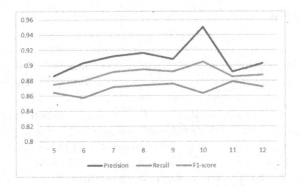

(a) Experimental Results on Metrics with Different K

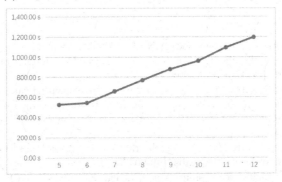

(b) Training Time with Different K

Fig. 5. Comparison results with different K

The parameter K controls the number of sub-classifiers of each primary classifier. In order to demonstrate the effect of K, we change the value of K at the range of $[5, 12]$ and the experimental results are shown in Fig. 5(a). We can observe that with the increase of K value, the performance climbs up at first and then declines. The best performance (i.e. $F_1score = 0.9050$) is achieved at $K = 10$ while the worst happens at $K = 5$ (i.e. $F_1score = 0.8752$). The reason behind this phenomenon is that when K is small, the classifier is greatly affected by overfitting and when K is large, the difference between the sub-classifiers is not obvious, so that the classification results cannot be distinguished. In addition, as K increases, the number of classifier training also increases. Figure 5(b) shows that the training time and K are approximately positively correlated. Therefore, it is recommended to set K at the range of $[8, 10]$.

5   Conclusion

Nowadays, detecting malicious attack in encrypted traffic is a great challenge. In this paper, we design an ensemble learning framework based on stack algorithms to identify encrypted malicious traffic. The addition of ensemble learning algorithms enables the method proposed in this paper to deeply characterize the behavior of the attacker and realize the effective identification of malicious traffic. Our experimental results demonstrated that our method achieves an F1-score of 0.905 and outperforms the state-of-the-art methods in real-world settings. In future work, we look forward to further improving its efficiency to accommodate the need of traffic classification on high-speed network.

References

1. Alshammari, R., Zincir-Heywood, A.N.: Investigating two different approaches for encrypted traffic classification. In: 2008 Sixth Annual Conference on Privacy, Security and Trust, pp. 156–166 (2008)
2. Anderson, B., McGrew, D.: Machine learning for encrypted malware traffic classification: accounting for noisy labels and non-stationarity. In: Proceedings of the 23rd ACM SIGKDD International Conference on Knowledge Discovery and Data Mining (2017)
3. Bahdanau, D., Cho, K., Bengio, Y.: Neural machine translation by jointly learning to align and translate. CoRR abs/1409.0473 (2015)
4. Breiman, L.: Stacked regressions. Mach. Learn. **24**, 49–64 (1996)
5. Breiman, L.: Random forests. Mach. Learn. **45**, 5–32 (2004)
6. Cao, Z., Xiong, G., Zhao, Y., Li, Z., Guo, L.: A survey on encrypted traffic classification (2014)
7. Chen, Y., Zang, T., Zhang, Y., Zhou, Y., Wang, Y.: Rethinking encrypted traffic classification: a multi-attribute associated fingerprint approach. In: 2019 IEEE 27th International Conference on Network Protocols (ICNP), pp. 1–11 (2019)
8. Qianxin Group and Tsinghua University: DataCon 2020. https://datacon.qianxin.com/opendata/maliciousstream. Accessed Aug 2020
9. Hochreiter, S., Schmidhuber, J.: Long short-term memory. Neural Comput. **9**, 1735–1780 (1997)
10. Li, R., Xiao, X., Ni, S., Zheng, H., Xia, S.: Byte segment neural network for network traffic classification. In: 2018 IEEE/ACM 26th International Symposium on Quality of Service (IWQoS), pp. 1–10 (2018)
11. Liu, C., He, L., Xiong, G., Cao, Z., Li, Z.: FS-Net: a flow sequence network for encrypted traffic classification. In: IEEE INFOCOM 2019 - IEEE Conference on Computer Communications, pp. 1171–1179 (2019)
12. Liu, C., Cao, Z., Xiong, G., Gou, G., Yiu, S., He, L.: MaMPF: encrypted traffic classification based on multi-attribute Markov probability fingerprints. In: 2018 IEEE/ACM 26th International Symposium on Quality of Service (IWQoS), pp. 1–10 (2018)
13. Lotfollahi, M., Jafari Siavoshani, M., Shirali Hossein Zade, R., Saberian, M.: Deep packet: a novel approach for encrypted traffic classification using deep learning. Soft. Comput. **24**(3), 1999–2012 (2019). https://doi.org/10.1007/s00500-019-04030-2

14. Melo, W., Lopes, P., Antonello, R., Fernandes, S., Sadok, D.: On the performance of DPI signature matching with dynamic priority. In: 2014 IEEE Symposium on Computers and Communications (ISCC), pp. 1–6 (2014)

15. MontazeriShatoori, M., Davidson, L., Kaur, G., Lashkari, A.H.: Detection of DoH tunnels using time-series classification of encrypted traffic. In: 2020 IEEE International Conference on Dependable, Autonomic and Secure Computing, International Conference on Pervasive Intelligence and Computing, International Conference on Cloud and Big Data Computing, International Conference on Cyber Science and Technology Congress (DASC/PiCom/CBDCom/CyberSciTech), pp. 63–70 (2020)

16. Pan, W., Cheng, G., Tang, Y.: WENC: HTTPS encrypted traffic classification using weighted ensemble learning and Markov chain. In: 2017 IEEE Trustcom/BigDataSE/ICESS, pp. 50–57 (2017)

17. Paszke, A., et al.: PyTorch: an imperative style, high-performance deep learning library. In: NeurIPS (2019)

18. Google Transparency Report: HTTPS encryption on the web. https://transparencyreport.google.com/https/overview?hl=en/

19. Shen, M., Zhang, J., Zhu, L., Xu, K., Du, X.: Accurate decentralized application identification via encrypted traffic analysis using graph neural networks. IEEE Trans. Inf. Forensics Secur. **16**, 2367–2380 (2021)

20. Shi, H., Li, H., Zhang, D., Cheng, C., Cao, X.: An efficient feature generation approach based on deep learning and feature selection techniques for traffic classification. Comput. Networks **132**, 81–98 (2018)

21. Su, J., Chen, S., Han, B., Xu, C., Wang, X.: A 60Gbps DPI prototype based on memory-centric FPGA. In: Proceedings of the 2016 ACM SIGCOMM Conference (2016)

22. Taylor, V.F., Spolaor, R., Conti, M., Martinovic, I.: AppScanner: automatic fingerprinting of smartphone apps from encrypted network traffic. In: 2016 IEEE European Symposium on Security and Privacy (EuroS&P), pp. 439–454 (2016)

23. Ting, K., Witten, I.: Issues in stacked generalization. J. Artif. Intell. Res. **10**, 271–289 (1999)

24. Velan, P., Cermk, M., Celeda, P., Drasar, M.: A survey of methods for encrypted traffic classification and analysis. Int. J. Netw. Manag. **25**, 355–374 (2015)

25. Xing, J., Wu, C.: Detecting anomalies in encrypted traffic via deep dictionary learning. In: IEEE INFOCOM 2020 - IEEE Conference on Computer Communications Workshops (INFOCOM WKSHPS), pp. 734–739 (2020)

26. Yao, H., Liu, C., Zhang, P., Wu, S., Jiang, C., Yu, S.: Identification of encrypted traffic through attention mechanism based long short term memory. IEEE Trans. Big Data, 1 (2019)

A Federated Learning Assisted Conditional Privacy Preserving Scheme for Vehicle Networks

Zhe Xia[1,2]([✉]), Yifeng Shu[1], Hua Shen[3], and Mingwu Zhang[3]

[1] School of Computer Science and Artificial Intelligence,
Wuhan University of Technology, Wuhan 430070, China
xiazhe@whut.edu.cn
[2] Guangxi Key Laboratory of Trusted Software, Guilin University of Electronic
Technology, Guilin 541004, China
[3] School of Computers, Hubei University of Technology, Wuhan 430068, China

Abstract. As the development of mobile communication technologies, Vehicle Networks can not only improve the efficiency of traffic operation, but also enhance the intelligent management level of traffic services. However, Vehicle Networks also bring a series of challenges, such as information leakage and message manipulation. In this paper, we introduce a novel federated learning assisted privacy preserving scheme for Vehicle Networks. In the proposed scheme, pseudonym is employed to hide the real identity of the vehicle, and homomorphic encryption is used to protect the private information in the training and aggregation processes. Moreover, the system is assisted with federated learning and fog computing. This not only improves efficiency in data integration and transmission, but also contributes to a more flexible and controllable traffic system. Security analyses demonstrate that the scheme meets the desirable security requirements, such as correctness, conditional privacy preserving and message authentication. And compared with some existing schemes, our proposed scheme enjoys better efficiency in both computation and communication.

1 Introduction

Recently, with the rapid development of information and communication technologies, Vehicle Networks have attracted great attentions both in academia and industry. It is an important technique to construct the Intelligent Traffic System (ITS). In general, Vehicle Network is a heterogeneous network composed of vehicles and infrastructures, such as roadside units (RSUs). The interior of each vehicle is a local area network. The vehicles form an inter-vehicle network, which is then connected to the Internet. All entities are based on a unified protocol to realise the integration of users, vehicles, roads, and clouds. The development of Vehicle Networks not only improves travel experience for the users, but also demonstrates good potentials for automatic driving in the future. It is widely recognised that it is a valuable technology for the green environment, thanks to its merits in traffic jam reduction and fuel consumption saving.

© Springer Nature Switzerland AG 2022
W. Meng and M. Conti (Eds.): CSS 2021, LNCS 13172, pp. 16–35, 2022.
https://doi.org/10.1007/978-3-030-94029-4_2

In the Vehicle Networks, all vehicles can periodically send travel information, such as speed, acceleration rate, location, weather, traffic conditions, etc. This traffic information can be processed and analysed to assist route planning for vehicles, enhance the stability and safety of the entire Vehicle Networks. Although data transmission in Vehicle Networks improves the traffic management level, it also brings a series of security issues [2]. For example, the information sent by vehicles not only contains the user's identity, but also may disclose some other sensitive information, such as the user's frequently visiting places. Adversaries can intercept this information through open wireless communication channels, posing a serious threat to user anonymity and privacy. Moreover, adversaries can also actively tamper the exchanged messages, affecting the authenticity of traffic information. Such attacks are more catastrophic, e.g. causing traffic accidents. Therefore, the anonymity of user identity and data security have become one of the main obstacles to the development and deployment of Vehicle Networks in the future.

Over the past few years, a number of privacy preserving and identity hiding schemes have been proposed in the literature, trying to protect the anonymity of vehicles. However, if vehicles' identities are protected in an unconditional way, it may also cause legal issues. For example, once some driver commits a crime, she cannot be traced and her misbehaviour is not accountable. Therefore, privacy preserving solutions for Vehicle Networks not only need to provide user anonymity, but also need to trace the malicious user's identity when necessary. This security requirement is called conditional privacy preserving [9], i.e. one can trace the real identity of vehicles under some specific conditions. Note that it is of great significance to consider this security requirement in real-world applications.

1.1   Our Contributions

To address the above problem and taking into account the efficiency requirement in Vehicle Networks, we propose a federated learning assisted conditional privacy preserving scheme for Vehicle Networks. The main contributions are as follows: 1) The integration of federated learning enables all vehicles to collect and process traffic data individually. Then, the local data sets are used to train and aggregate the Vehicle Networks models. Finally, these models are uploaded to a management center for analysis in order to improve the efficiency of data transmission and system operation. 2) Fog computing is also used in the system design, with the purpose of reducing the workload of the management center as well as improving the flexibility and efficiency of system operation. 3) Pseudonyms are employed to guarantee the anonymity of the vehicles, and homomorphic encryption is used to encrypt the trained models, protecting their secrecy from the others. In order to resist the collusion attack between the fog nodes, blinding technologies are used to hide the encryption results. Security and efficiency analyses show that the proposed scheme can not only achieve the desirable security requirements, such as correctness, conditional privacy preserving and message authentication, but also greatly improve the efficiency compared with the related schemes in the literature. Hence, it achieves a good balance between security and efficiency, and it is suitable for the resource constrained environment such as Vehicle Networks.

1.2   Organisation of the Paper

The rest of the paper is organised as follows: In Sect. 2, we review the existing privacy preserving schemes for Vehicle Networks. In Sect. 3, we outline some preliminaries. The models and definitions for the proposed scheme are described in Sect. 4. In Sect. 5, we introduce our proposed conditional privacy preserving scheme, and its security and efficiency analyses are presented in Sect. 6. Finally, we conclude in Sect. 7.

2   Related Works

Recently, privacy preserving in Vehicle Networks has become a popular research area. In general, the existing schemes can be classified as follows.

The first approach uses certificates in public key cryptography. In [6], the certificate authority issues the serial number and the certificate to every vehicle, while the receiver authenticates the vehicle's identity through the certificate. In [18], Raya et al. realised the authentication and integrity by pre-loading the public/private key pair and the corresponding certificate into the vehicle's on-board unit. But its limitation is that when there are a large number of users, the central node needs to store and manage too many certificates. Lu et al. [15] required the vehicles to request temporary anonymous certificates from the RSUs when passing through them, causing frequent interactions between the vehicles and the RSUs.

The second approach uses identity-based signature (IBS) to design privacy preserving scheme. The user's identity is the user's public key. Therefore, it does not require storage or management of the certificates, addressing the limitation in the certificate-based schemes [20]. The scheme proposed by Kamat et al. [11] can effectively simplify the certificate management process, reduce the burden of the central node, and reduce the communication overheads by using identity-based signature. In [1], Bayat et al. used an identity-based public key encryption to construct a conditional privacy preserving scheme. The user's identity is her public key, so one does not need a certificate to bind the user to her public key. However, the schemes in this category are vulnerable to replay attacks, simulation attacks, and their computational costs are high.

The third approach uses group signature to design privacy preserving schemes, in which the group members can sign messages anonymously, and the group managers can verify the identity of signers when necessary [27]. Liu et al. [13] used signature with special property to preserve privacy. Vehicles use the technology of non-certificate signing to obtain the key. The key was managed in partition, which reduces the key management center's burden and also solves the recall problem of the illegal vehicles. In [19], Shao et al. proposed a threshold anonymous authentication protocol through a group signature scheme. It integrates the decentralized group model and the threshold authentication method. Hence, it meets the security requirements of threshold authentication and privacy preserving. However, Shao's scheme is vulnerable to the collusion attack, where the betrayed group members can collude with the external attackers to disclose the privacy information of group members.

The fourth approach uses pseudonyms to preserve privacy of the vehicle's identity. In these schemes, pseudonyms are associated with the real identity of vehicles, and pseudonyms are used to cover up the real identity of vehicles, so that anonymity and non-traceability can be achieved. Palanisamy et al. [17] and Lu et al. [14] require the vehicles to update the pseudonyms in busy areas, increasing the uncertainty of pseudonyms and preventing pseudonyms from being tracked by adversaries over a long period. The limitation is that the privacy protection level maybe low in areas with only a fewer vehicles. In [23], Yu et al. constructed an extended group area for pseudonyms exchange, which enables to achieve a good privacy preserving effect even under the condition of sparse traffic. However, the security of the scheme depends on the assumption that the contents of network communication are not tampered with.

Traditional privacy preserving schemes for Vehicle Networks use cryptographic primitives, such as elliptic curve, symmetric encryption and hash function to preserve data privacy and identity privacy of the vehicles. Although strong anonymity guarantee enables the protection of vehicle's privacy information, malicious parties can also make use of the privacy preserving mechanism to hide their identity from being traced. Therefore, apart from protecting vehicle's anonymity, it is also required to provide the ability of tracing the vehicle in real-world applications. And this is important for controllable supervision of vehicle's identity. Such a requirement is called conditional privacy preserving.

In 2015, He et al. [8] proposed a conditional privacy preserving scheme without bilinear pairing and scalar multiplication, dramatically reducing the computational costs. Moreover, the batch verification technique is used to verify a large number of signatures in batch, which makes the scheme suitable for deployment in the resource constrained environment, such as Vehicle Networks. In 2018, Li et al. [12] proposed a provable secure conditional privacy preserving scheme, and this scheme enjoys low computation and communication overheads. Afterwards, Zhang et al. [24] proposed a new conditional privacy preserving authentication scheme based on Chinese Remainder Theorem. Although this scheme employs tamper proof devices, it does not require to pre-load the master key to each vehicle's tamper proof device, and it is secure against the side channel attacks.

Currently, various schemes have considered the requirements of protecting data transmission between vehicles and the anonymity of vehicles. However, if complex cryptographic primitives are employed, it will have a negative effect on the system performance, making it unlikely to meet the requirements in real-world applications. This is especially true for Vehicle Networks with limited resources. Therefore, to design privacy preserving solutions for Vehicle Networks, one should also consider the balance between security and efficiency.

3   Preliminaries

3.1   Federated Learning

Machine learning needs large amount of data to train the model. However, it is very difficult to integrate the data scattered in various places and institutions because of the phenomenon of data island. And meanwhile, data protection has been gradually emphasised. The security and privacy of user data has become more and more strict, making data sharing very difficult. Federated learning has emerged as the key technology to solve this industry problem.

Federated learning is a type of machine learning in the distributed setting. Coordinated by the central server, many clients work together to train the model, while keeping the training data decentralized. Its goal is to utilise the data from multiple users while maintaining data privacy, security and legitimacy, and finally obtain a global model [22]. According to the distributed characteristics of island data, federated learning can be divided into three categories: horizontal one, vertical one and federated transfer learning [25], as shown in Fig. 1.

Fig. 1. Analytical diagram of federated learning

Federated learning has been found a useful technique in Vehicle Networks, in which different vehicles in different regions are continuously producing a large amount of traffic data. This paper utilises horizontal federal learning to improve the efficiency in information transmission, and it also reduces the workload of trusted authority in Vehicle Networks. For example, one can integrate a large amount of data with the same data characteristics in different regions, and then train and optimise the Vehicle Networks models using the local data sets. As follows, the system carries out real-time management, that benefits all vehicles in the Vehicle Networks, such as improving the efficiency of data transmission as well as maintaining the stable operation of the transportation system [16].

3.2   Fog Computing

Fog computing is a natural extension of cloud computing, by adding a layer of "fog" between the terminal equipments and the cloud server. That is, the network edge layer only concentrates the user data, but not the data stored in the cloud. This can greatly reduce the computation and storage overheads in the cloud, improve the transmission rate, and reduce the delay. Moreover, its architecture is designed in the distributed fashion. And this enables distributed packet transmission and distributed data control, reducing the possibility of network congestion [4]. The architecture of fog computing is shown in Fig. 2.

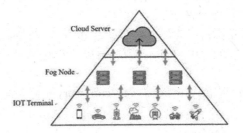

Fig. 2. Architecture of fog computing

It is obvious that the architecture of "cloud server - fog node - IoT terminal" is very similar to the system model of "trusted authority - roadside unit - vehicle" in Vehicle Networks. Unlike the cloud server, fog nodes do not need powerful performance to integrate and manage a large number of data resources. They are usually composed of various functional computers with limited computation and storage capacity. The roadside unit in Vehicle Networks manages the collection of vehicle information in a specific area, but it does not need to process and analyse the data. Hence, it does not need to have strong performance. At the same time, fog nodes have many characteristics, such as low delay, location awareness, wide geographical distribution, etc. Roadside unit is also widely deployed by the road, which is time sensitive, more reliable and convenient [21]. Therefore, this paper uses the fog computing technology to outsource the workload of aggregating and processing the models by the trusted authority to the cloud server. The roadside units are only treated as the fog nodes to collect the vehicle data in its management area, train models and send them to the cloud server. Compared with the traditional Vehicle Networks, this method not only reduces the workload for both trusted authority and the roadside units, but also improves the data transmission rate and the efficiency of system operation.

3.3 Homomorphic Encryption

With the development of cloud computing and artificial intelligence, homomorphic encryption becomes a very popular tool. Compared with the traditional encryption algorithms, homomorphic encryption can not only realise the basic encryption operation, but also satisfies some attractive computational functions between ciphertexts. That is, ciphertexts can be computed before decrypting, and the function is equivalent to computing the plaintexts after decrypting.

In this paper, we use the homomorphic encryption scheme introduced in [3] with the following homomorphic property: $Enc(m_1) \cdot Enc(m_2) = Enc(m_1 + m_2)$.

The purpose is to prevent malicious cloud servers from learning the models in federated learning. The main algorithms are as follows:

Key Generation. Denote $N = pq$, where p, q are large primes. Choose a random element $\alpha \in \mathbb{Z}_{N^2}^*$, a random value $a \in [1, \mathrm{ord}(\mathbb{G})]$, and set $g = \alpha^2 \bmod N^2$ and $h = g^a \bmod N^2$. The public key is given by the triplet (N, g, h) while the corresponding secret key is a.

Encrypt. Given a message $m \in \mathbb{Z}_N$, a random pad r is chosen uniformly and at random in $\mathbb{Z}_{N^2}$, the ciphertext (A, B) is computed as $A = g^r \bmod N^2$, $B = h^r(1 + mN) \bmod N^2$.

Decrypt. Knowing the private key a, one can decrypt as $m = \frac{B/(A^a) - 1 \bmod N^2}{N}$.

4   Models and Definitions

4.1   System Model

In the system architecture, each vehicle is used as terminal to collect real-time traffic data in the process of driving, and it uploads the traffic data to the roadside units, which are treated as the fog nodes in the system. It trains a local model with the received vehicles' data, encrypts and blinds the trained local model to preserve its privacy, and uploads the updated local model in the current round to cloud server. The cloud server then aggregates the received data and generates a new round of global model. It then transmits the encrypted global model to the trusted authority, who will decrypt and analyse the global model, and make management decisions for the system. Note that the system model assigns the workload of training and aggregating the traffic data to the fog node and the cloud server, so that the workload of the trusted authority can be greatly reduced. The communication channels between vehicles and fog nodes, and between fog nodes and the cloud server are wireless, using dedicated short range communication (DSRC) protocol. While the communication between fog node and the trusted authority is exchanged through the secure wired channel. Our proposed system model for Vehicle Networks is shown in Fig. 3.

- **Trusted Authority (TA):** TA (e.g. Vehicle Management Bureau) is the only trustworthy entity in the system model. It is assumed to have large computation and storage capacity and can handle large and complex data models. It is mainly responsible for the registration of system entities, the generation of system parameters, the generation of pseudonyms, the analysis of models, and the distribution of decision-making.
- **Cloud Server (CS):** To reduce TA's workloads, CS can perform some computational tasks instead, e.g. it aggregates all the local models uploaded by fog nodes to get the global model, and then sends it to TA for processing. But it is assumed that the CS is honest-but-curious.

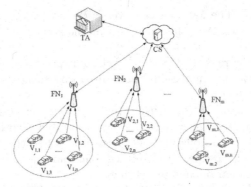

Fig. 3. System model of Vehicle Networks

- **Fog Node (FN):** FN is the roadside unit deployed at road intersections or by the road. It is responsible for collecting all vehicles' data in an area and training new local model using these data sets. Moreover, it generates blind parameters, and blinds the new local model. FN is also responsible to distribute the system parameters and decisions to the vehicles.
- **Vehicles (V):** Vehicles are users in the system. They collect information by sensors. The onboard unit (OBU) is used as a communication tool to communicate with FN, and upload the collected traffic data to the FN nearby.

4.2   Adversary Model

The adversary can attack the Vehicle Networks in different ways. For example, it may try to find out the vehicle's identity, or it may intend to learn the models within the Vehicle Networks. In addition, it may try to tamper the models, resulting in system collapse and traffic disorder. Therefore, this paper considers two types of adversaries, external ones and internal ones.

- **External adversary:** The external adversary refers to the malicious attackers outside the Vehicle Networks, who can not only eavesdrop the messages sent through the channel with the purpose of learning the vehicle's identity or the local models, but also launch active attacks, e.g. tamper, forge, delete, replay the messages, destroy the integrity of messages.
- **Internal adversary:** Internal adversary refers to the dishonest internal entities of Vehicle Networks. They may leak their local models to the others. Note that in the proposed system model, only the trusted authority is assumed to be honest, while the others are all honest-but-curious.

4.3   Security Requirements

Security and privacy are very important for the communication of Vehicle Networks, and various security challenges have been extensively studied. In general, a conditional privacy preserving scheme should meet the following security requirements.

- **Data Integrity:** The fog nodes can verify the authenticity of the messages sent by the vehicles. And this ensures that the received traffic data is valid and not tampered.
- **Anonymity:** The real identity of the vehicle is preserved, and no adversary should be able to obtain this information by observing the messages transmitted by the vehicles.
- **Traceability:** In order to prevent the malicious vehicles from using the anonymity mechanism in a harmful way, their real identity can be recovered when some conditions are met. Note that when the above anonymity and traceability properties are achieved simultaneously, it is said that the system is conditional privacy preserving.
- **Model Confidentiality:** The global model of the Vehicle Networks should be only known by the trusted authority. Moreover, the local models by honest fog nodes should not be known by the others.

5   The Proposed Scheme

The proposed scheme contains five algorithms: *Initialization, Vehicle Data Collection, Fog Node Training, Cloud Server Aggregating, Model Analysis and Regulation*. In the system, suppose there exists a TA, a CS, m FNs and each FN manages n vehicles. At first, TA generates and announces the system parameters. Then, each entity registers with TA, and TA generates the initial model of Vehicle Networks. In each round, the vehicles first collect the traffic information data and send it to FN, who then downloads the global model of the previous round from CS, trains the model with the data set to generate a new local model, and sends the local model to CS after blinding and encryption. CS aggregates all the local models to generate a new global model and sends it to TA, who then decrypts and analyses the global model. Then it enters the next round to maintain the stable operation of the system. The flow chart of system operation is shown in Fig. 4.

5.1   Initialization

Initialization phase mainly includes three parts: entity registration, generation of system parameters and key distribution, and generation of the initial model of Vehicle Networks. The detailed steps of each part are as follows:

1. The cloud server CS, all fog nodes $FN_i(i = 1, 2, ..., m)$ and all vehicles $V_{i,j}(j = 1, 2, ..., n)$ send registration request to the TA. TA responses according to the legitimacy of each entity. The registration response of the legitimate vehicle should include the real identity $RID_{i,j}(j = 1, 2, ..., n)$ generated by TA as the unique identifier of the vehicle.
2. TA chooses two large primes p, q and a non-singular elliptic curve E defined by the equation $y^2 = x^3 + ax + b \bmod p$, where $a, b \in \mathbb{F}_p$; And it chooses a generator P with order q of the group $\mathbb{G}$, where $\mathbb{G}$ consists of all points on

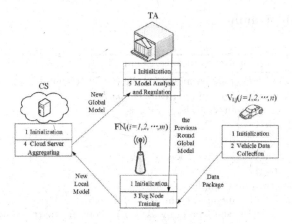

Fig. 4. Flow chart of system operation

the elliptic curve E and the point at infinity O. Then, TA chooses a random number $x \in \mathbb{Z}_q^*$ as the private key of the system and the system public key is $P_{pub} = x \cdot P$; Next, TA chooses three secure hash functions h_1, h_2, h_3, where $h_1 : \mathbb{G} \to \mathbb{Z}_q$, $h_2 : \{0,1\}^* \to \mathbb{Z}_q$, $h_3 : \{0,1\}^* \times \{0,1\}^* \times \mathbb{G} \times \{0,1\}^* \to \mathbb{Z}_q$. Finally, TA announces the system parameters $(P, a, b, P_{pub}, h_1, h_2, h_3, N, g, h)$. The key pairs (N, g, h, a) of the homomorphic encryption are distributed to each fog node $\text{FN}_i (i = 1, 2, ..., m)$ through a secure channel.

3. TA downloads the model from the model provider or collects the data set from the legitimate users to generate the initial model Ω_0. It then encrypts it as $[\Omega_0]_{pk}$ and sends the encrypted initial model to CS.

5.2   Vehicle Data Collection

Vehicle Data Collection phase includes pseudonym generation, traffic information collection and data package generation. The detailed steps are as follows:

1. Vehicle $V_{i,j}$ generates the pseudonym periodically according to the real identity $RID_{i,j}$. First, it generates a random number $r_{i,j} \in \mathbb{Z}_q^*$ and computes $AID_{i,j,1} = r_{i,j} \cdot P$, $AID_{i,j,2} = RID_{i,j} \oplus h_1(r_{i,j} \cdot P_{pub})$. The vehicle pseudonym is $AID_{i,j} = (AID_{i,j,1}, AID_{i,j,2})$. Then $V_{i,j}$ computes $\alpha_{i,j} = h_2(AID_{i,j}||T_{i,j})$, $\rho_{i,j} = r_{i,j} + \alpha_{i,j} \cdot x \ mod \ q$, where $T_{i,j}$ is the current timestamp, $\oplus$ is XOR operation, $||$ stands for concatenation, $\alpha_{i,j}$ is the hash value of hash function h_2, $\rho_{i,j}$ denotes a Schnorr Signature on $\alpha_{i,j}$.

2. $V_{i,j}$ collects real-time traffic information $M_{i,j}$. Then, $V_{i,j}$ generates data package using its pseudonym $AID_{i,j}$ and $M_{i,j}$. It first generates a random number $k_{i,j} \in \mathbb{Z}_q^*$, and then computes

$$K_{i,j} = k_{i,j} \cdot P, \beta_{i,j} = h_3(AID_{i,j}||T_{i,j}||K_{i,j}||M_{i,j}), \sigma_{i,j} = \rho_{i,j} + \beta_{i,j} \cdot k_{i,j} \ mod \ q$$

where $\beta_{i,j}$ is the hash value of hash function h_3, $\sigma_{i,j}$ is the Schnorr Signature for this value. $V_{i,j}$ then sends $DP_{i,j} = \{M_{i,j}, AID_{i,j}, T_{i,j}, K_{i,j}, \sigma_{i,j}\}$ to FN_i.

5.3 Fog Node Training

Algorithm 1. Training and Generate Local Model in Fog Node

Input: data package $DP_{i,j}$, global model for the $(t-1)$-th round $[\Omega_{t-1}]_{pk}$, the number of fog nodes m, upper limit of training rounds *epoch*.
Output: blind encrypted local model $\eta_{i,t} * [\omega_{i,t}]_{pk}$.
for $i \leq epoch$ FN$_i$ **do**

1: Batch verification of all data packages $DP_{i,j}$.
2: Put all $M_{i,j}$ into the local data set.
3: Generate the t-th round blinding factor $\eta_{i,t}$.
4: Download the $(t-1)$-th round ciphertext of global model $[\Omega_{t-1}]_{pk}$ from CS.
5: Decrypt $[\Omega_{t-1}]_{pk}$ using the secret key $sk(\lambda, \mu)$ to obtain the $(t-1)$-th round global model Ω_{t-1}.
6: Get a mini-batch of data from its local data set.
7: Train and generate local model $\omega_{i,t}$ by using the BP algorithm based on SGD.
8: Encrypt $\omega_{i,t}$ using public key $pk(N, g)$ to obtain the ciphertext $[\omega_{i,t}]_{pk}$.
9: Blind $[\omega_{i,t}]_{pk}$ using the blinding factor $\eta_{i,t}$ to obtain $\eta_{i,t} * [\omega_{i,t}]_{pk}$.
10: Upload the blind encrypted local model $\eta_{i,t} * [\omega_{i,t}]_{pk}$ to CS.
end for

In the proposed scheme, FN$_i$ trains and generates the encrypted blind local model according to Algorithm 1. The first step of the algorithm indicates that FN$_i$ will verify all the received vehicle data packages $DP_{i,j}$ in t-th round. In order to improve the efficiency of verification, the scheme uses small index test technology in [10] to batch verify n data packages to ensure that the received packages are all from legitimate vehicles in the system and the packages have not been tampered or forged. First, FN$_i$ will check the freshness of the timestamps $T_{i,j}$ in the data package. If not, FN$_i$ will reject the package. Then, FN$_i$ accepts all the fresh packages and randomly generates a set of vectors $\{v_{i,1}, v_{i,2}, ..., v_{i,n}\}$, where $v_{i,j}$ is a small random integer in $[1, 2^t]$ and t is a small integer. Then we check whether the following equation holds:

$$(\sum_{j=1}^{n} v_{i,j} \cdot \sigma_{i,j}) \cdot P = (\sum_{j=1}^{n} v_{i,j} \cdot AID_{i,j,1}) + (\sum_{j=1}^{n} (v_{i,j} \cdot \alpha_{i,j})) \cdot P_{pub} + \sum_{j=1}^{n} (v_{i,j} \cdot \beta_{i,j} \cdot K_{i,j})$$

$$(1)$$

If it holds, all data packages $DP_{i,j}$ within the scope of FN$_i$ management are considered valid, and then FN$_i$ adds the traffic information data $M_{i,j}$ to the local data set.

In the third step of the algorithm, FN$_i$ begins to generate the current round of blinding factor, which is based on a two round anonymous veto protocol in [5]. In the first round, each participant randomly selects $x_i \in Z_q$, then publishes g^{x_i}, where g is the generator of a cyclic group of prime order q; After this round, each participant in the second round can calculate $g^{y_i} = \prod_{j=1}^{i-1} g^{x_j} / \prod_{j=i+1}^{n} g^{x_j}$

according to the information in the first round. Therefore, each FN_i can calculate its blinding factor $\eta_{i,t} = g^{x_i}g^{y_i}$ in t-th round.

In steps 4 and 5 of the algorithm, FN_i obtains the global model of the last round. FN_i first downloads the ciphertext $[\Omega_{t-1}]_{pk}$ of the global model in $(t-1)$-th round from CS (if it is the first round, download the initial model ciphertext $[\Omega_0]_{pk}$), and decrypts it with the **Decryption** algorithm in Sect. 3.3 to obtain the global model Ω_{t-1} of $(t-1)$-th round.

The sixth and seventh steps of the algorithm are two steps of training the model. FN_i will get a small batch of data from the local data set to calculate the gradient, and use the Back-Propagation (BP) algorithm based on stochastic gradient descent (SGD) [26] to train the model:

$$\omega_{i,t} = \frac{\Omega_{t-1}}{m} - \alpha \frac{\delta E_m^i}{\delta \frac{\Omega_{t-1}}{m}} \tag{2}$$

where α is the learning rate, E is the error function, and the local model of the t-th round generated by FN_i is $\omega_{i,t}$.

In the last three steps, FN_i encrypts and blinds the local model of t-th round and uploads it to CS. First, it uses **Encryption** algorithm in Sect. 3.3 to encrypt global model $\omega_{i,t}$ and get the ciphertext $[\omega_{i,t}]_{pk}$ of it, and then use the blinding factor $\eta_{i,t}$ obtained in step 5 to blind the ciphertext $[\omega_{i,t}]_{pk}$, finally FN_i computes the blind encrypted local model $\eta_{i,t} * [\omega_{i,t}]_{pk}$ and upload it to CS.

5.4   Cloud Server Aggregating

Algorithm 2. Aggregation and generate global model in Cloud Server

---

Input: blind encrypted local model $\eta_{i,t} * [\omega_{i,t}]_{pk}$ of all FNs, upper limit of training rounds *epoch*.
Output: encrypted global model $[\Omega_t]_{pk}$.
for $t \le epoch$ CS **do**

    1: Receive $\eta_{i,t} * [\omega_{i,t}]_{pk}$ of all fog nodes.
    2: Aggregate all the local models of the Vehicle Networks system.
    3: Generate the ciphertext of global model $[\Omega_t]_{pk}$.
    4: Upload the encrypted global model $[\Omega_t]_{pk}$ to TA.
    5: Iteration round $t = t + 1$.
end for

In the proposed scheme, CS aggregates local models according to Algorithm 2 and generates the encrypted global model. First, CS receives the blind encrypted local models $\eta_{i,t} * [\omega_{i,t}]_{pk}$ transmitted by all fog nodes, and then use the following formula to aggregate them:

$$\eta_{1,t} * [\omega_{1,t}]_{pk} \cdot \eta_{2,t} * [\omega_{2,t}]_{pk} \cdot \ldots \cdot \eta_{m,t} * [\omega_{m,t}]_{pk} = \prod_{i=1}^{m} \eta_{i,t} \cdot \prod_{i=1}^{m} [\omega_{i,t}]_{pk} \qquad (3)$$

Due to the characteristics of the blind technology based on a two round anonymous veto protocol in [27], $\sum_{i=1}^{n} x_i y_i = \sum_{i=1}^{n} x_i (\sum_{j=1}^{i-1} x_j - \sum_{j=i+1}^{n} x_j) = 0$, we can compute $\prod_{i=1}^{m} \eta_{i,t} = \prod_{i=1}^{m} g^{x_i y_i} = g^{\sum_{i=1}^{m} x_i y_i} = g^0 = 1$, the operation of deblinding can be realized. And because of the homomorphic property $Dec(Enc(m_1) \cdot Enc(m_2)) = m_1 + m_2$ of homomorphic encryption scheme in [3], we compute $\prod_{i=1}^{m} [\omega_{i,t}]_{pk} = [\sum_{i=1}^{m} \omega_{i,t}]_{pk}$. So CS can generate the un-blinded encrypted global model $[\Omega_t]_{pk} = \prod_{i=1}^{m} \eta_{i,t} \cdot \prod_{i=1}^{m} [\omega_{i,t}]_{pk} = [\sum_{i=1}^{m} \omega_{i,t}]_{pk}$ from the above conditions. Finally, CS sends $[\Omega_t]_{pk}$ to TA.

5.5   Model Analysis and Regulation

After receiving the un-blinded encrypted global model $[\Omega_t]_{pk}$ sent by CS, TA uses the **Decrypt** algorithm in Sect. 3.3 to decrypt it, and obtains the global model Ω_t of the t-th round. At this time, TA can analyse the global model, carry out macro-control on the Vehicle Networks system according to the analysis results, and implement specific measures according to the situation of different regions, so as to maintain the stable operation of the whole system.

6   Security and Efficiency Analysis

In this section, according to the adversary model mentioned above, we consider various attacks the adversary may take and analyse the security of the proposed scheme. Moreover, the performance of the proposed scheme is evaluated by the computation costs and communication costs, and it is compared with the existing conditional privacy preserving schemes.

6.1   Security Analysis

Theorem 1 (Data Integrity). *The scheme has unforgeability under adaptive chosen message attacks in the random oracle model.*

Proof. This security property can be modeled by a game between challenger $\mathcal{C}$ and adversary $\mathcal{A}$. Suppose $\mathcal{A}$ is a probabilistic polynomial time adversary. He has the authority to control the communications within the Vehicle Networks. For example, $\mathcal{A}$ can monitor, modify or even forge messages. Suppose $\mathcal{C}$ can run $\mathcal{A}$ as a subroutine to solve the elliptic curve discrete logarithm problem (ECDLP) with non-negligible probability, we show that $\mathcal{C}$ can simulate $\mathcal{A}$'s query in the random oracle model. The details are as follows:

Setup: $\mathcal{C}$ generates the private key x as well as the system parameters, computes the corresponding public key $P_{pub} = x \cdot P$, and then sends the system parameters $parmas = \{p, q, a, b, P, P_{pub}, h_1, h_2, h_3\}$ to A.

h_1 *Query:* $\mathcal{C}$ generates a list L_{h1} with the form of (Γ, τ), and it is initialized to empty. After receiving $\mathcal{A}$'s query with the message Γ, $\mathcal{C}$ will check whether a tuple (Γ, τ) exists in L_{h1}. If not, $\mathcal{C}$ will generate a random number $\tau \in \mathbb{Z}_q$, add tuple (Γ, τ) to L_{h1}, and send $\tau = h_1(\Gamma)$ to $\mathcal{A}$; If so, $\mathcal{C}$ sends $\tau = h_1(\Gamma)$ to $\mathcal{A}$ directly.

h_2 *Query:* $\mathcal{C}$ generates a list L_{h2} with the form of (AID_i, T_i, τ), and it is initialized as empty. After receiving $\mathcal{A}$'s query with the message (AID_i, T_i), $\mathcal{C}$ checks whether the tuple (AID_i, T_i, τ) already exists in L_{h2}. If not, $\mathcal{C}$ selects a random number $\tau \in \mathbb{Z}_q$, adds the tuple (AID_i, T_i, τ) into the list L_{h2}, and sends $\tau = h_2(AID_i||T_i)$ to $\mathcal{A}$. Otherwise, $\mathcal{C}$ sends $\tau = h_2(AID_i||T_i)$ to $\mathcal{A}$ directly.

h_3 *Query:* $\mathcal{C}$ generates a list L_{h3} with the form of $(AID_i, T_i, K_i, M_i, \tau)$, and it is initialized as empty. After receiving $\mathcal{A}$'s query with the message (AID_i, T_i, K_i, M_i), $\mathcal{C}$ checks whether a tuple $(AID_i, T_i, K_i, M_i, \tau)$ already exists in L_{h3}. If not, $\mathcal{C}$ generates a random number $\tau \in \mathbb{Z}_q$, adds the tuple $(AID_i, T_i, K_i, M_i, \tau)$ into the list L_{h3}, and sends $\tau = h_3(AID_i||T_i||K_i||M_i)$ to $\mathcal{A}$. Otherwise, $\mathcal{C}$ sends $\tau = h_3(AID_i||T_i||K_i||M_i)$ to $\mathcal{A}$ directly.

Sign Query: After receiving $\mathcal{A}$'s query with the message M_i, $\mathcal{C}$ generates three random numbers $\alpha, \beta, \sigma \in \mathbb{Z}_q^*$, and chooses a random point $AID_{i,2}$, computes $AID_{i,1} = \sigma_i \cdot P - \alpha_i \cdot P_{pub} - \beta_i \cdot K_i$, and then adds the two tuples (AID_i, T_i, α_i) and $(AID_i, T_i, K_i, M_i, \beta_i)$ into the lists L_{h2} and L_{h3}, respectively, where $AID_i = \{AID_{i,1}, AID_{i,2}\}$. Finally, $\mathcal{C}$ sends the message $(M_i, AID_i, T_i, K_i, \sigma_i)$ to $\mathcal{A}$. At this moment, it can be verified that the equation $\sigma_i \cdot P = AID_{i,1} + \alpha_i \cdot P_{pub} + \beta_i \cdot K_i$ holds. Therefore, the above simulation is indistinguishable from the real scheme.

Equipped with the abilities modeled by the above oracles, $\mathcal{A}$ outputs a message $(M_i, AID_i, T_i, K_i, \sigma_i)$. Now, $\mathcal{C}$ can verify the validity of the message by the following equation:

$$\sigma_i \cdot P = AID_{i,1} + \alpha_i \cdot P_{pub} + \beta_i \cdot K_i \tag{4}$$

According to the forking lemma [7], if $\mathcal{A}$ can output one valid message, he also has a non-negligible probability to select a different output for h_2 *Query* to repeat the above process, and output another valid message $(M_i, AID_i, T_i, K_i, \sigma_i')$. It also can be verified by the following equation.

$$\sigma_i' \cdot P = AID_{i,1} + \alpha_i' \cdot P_{pub} + \beta_i \cdot K_i \tag{5}$$

According to the Eqs. (4) and (5), we can get $(\sigma_i - \sigma_i') \cdot P = (\alpha_i - \alpha_i') \cdot P_{pub} = (\alpha_i - \alpha_i') \cdot x \cdot P$, and $(\sigma_i - \sigma_i') = (\alpha_i - \alpha_i') \cdot x$. Therefore, $\mathcal{C}$ can output $(\alpha_i - \alpha_i')^{-1}(\sigma_i - \sigma_i')$ as the answer of the ECDLP. But it contradicts the hardness of ECDLP. So we can conclude that the proposed scheme has the unforgeability property under adaptive chosen message attacks in the random oracle model.

Theorem 2 (Anonymity and Traceability). *In the proposed scheme, the identity information of the vehicle meets the conditional privacy preserving.*

Proof. Conditional privacy preserving means that in the normal cases, the vehicle's identity information is preserved and it cannot be obtained by the adversary.

But in some emergence cases, some trusted third party in the system can track the real identity of the vehicle when some conditions are met. The purpose is to prevent the vehicle from mis-using the anonymity mechanism to conduct crimes.

In the proposed scheme, the real identity of the vehicle is created by TA as its unique identifier when the vehicle is registered, and the vehicle will generate its own pseudonym $AID_{i,j} = (AID_{i,j,1}, AID_{i,j,2})$ before collecting the traffic data, where $AID_{i,j,1} = r_{i,j} \cdot P$ and $AID_{i,j,2} = RID_{i,j} \oplus h_1(r_{i,j} \cdot P_{pub})$. So the adversary needs to compute $RID_{i,j} = AID_{i,j,2} \oplus h_1(r_{i,j} \cdot P_{pub})$ to obtain the vehicle's real identity. But the relevant conditions adversary can only get are $r_{i,j} \cdot P_{pub} = r_{i,j} \cdot x \cdot P$ and $AID_{i,j,1} = r_{i,j} \cdot P$. Therefore, the adversary has to solve the computational Diffie-Hellman (CDH) problem. Because the CDH problem is assumed to be hard, the adversary can not calculate the real identity of the vehicle. For the trusted TA in the system, it can compute $RID_{i,j} = AID_{i,j,2} \oplus h_1(r_{i,j} \cdot P_{pub})$ to obtain the real identity of the vehicle. The above analysis shows that the scheme meets the conditional privacy preserving for vehicle identity information.

Theorem 3 (Model Confidentiality). *The aggregated Vehicle Networks model of this scheme satisfies confidentiality for all entities except the trusted authority.*

Proof. For the global model generated by the cloud server, its confidentiality should be guaranteed. Regarding the individual model collected by each note, it is always transmitted in ciphertext using homomorphic cryptosystem in [3]. Because the encryption algorithm in [3] satisfies semantic security, it is impossible for the external adversary to obtain any information about the ciphertext without the private key. For the internal adversary, due to the characteristics of federated learning model, the security of outsourced cloud server and fog nodes is not guaranteed. Therefore, the proposed scheme considers that they are all semi-honest entities to prevent them from obtaining real-time global models. For the cloud server, the local models are all encrypted using homomorphic cryptosystem. Because the cloud server does not know the private key, it can not decrypt the encrypted global model after aggregating. Moreover, due to the characteristic of homomorphic encryption, cloud server can still calculate the ciphertext of the local models to complete the aggregation operation and keep the aggregated model confidential. For fog nodes, they can only get the local model of its own management area. Moreover, the scheme can resist the collusion attack between fog nodes. Suppose that there are $k(k \leq m - 1)$ fog nodes colluding, they want to aggregate models using their blind encrypted local models. But all the information that the colluding fog nodes can get is $\eta_{i,t} * [\omega_{i,t}]_{pk} (i = 1, ..., k \leq m - 1)$. If they cannot get all the blind encrypted local models, according to $\prod_{i=1}^{m} \eta_{i,t} = \prod_{i=1}^{m} g^{x_i y_i} = g^{\sum_{i=1}^{m} x_i y_i} = g^0 = 1$, the blind factor cannot be cancelled with each other, so the colluding fog nodes can not get any information about the global model. Through the above analysis, the global model of this scheme satisfies confidentiality for all entities except the trusted authority.

Theorem 4. *If formula (1) of the scheme is true, all data packages received by the fog nodes are valid.*

Proof. The correctness of formula (1) is deduced as follows:

$$(\sum_{j=1}^{n} v_{i,j} \cdot \sigma_{i,j}) \cdot P = (\sum_{j=1}^{n} v_{i,j} \cdot (\rho_{i,j} + \beta_{i,j} \cdot k_{i,j})) \cdot P$$

$$= (\sum_{j=1}^{n} v_{i,j} \cdot (r_{i,j} + \alpha_{i,j} \cdot x + \beta_{i,j} \cdot k_{i,j})) \cdot P$$

$$= (\sum_{j=1}^{n} v_{i,j} \cdot (r_{i,j} \cdot P + \alpha_{i,j} \cdot x \cdot P + \beta_{i,j} \cdot k_{i,j} \cdot P))$$

$$= (\sum_{j=1}^{n} (v_{i,j} \cdot AID_{i,j,1} + v_{i,j} \cdot \alpha_{i,j} \cdot P_{pub} + v_{i,j} \cdot \beta_{i,j} \cdot K_{i,j}))$$

$$= (\sum_{j=1}^{n} v_{i,j} \cdot AID_{i,j,1}) + (\sum_{j=1}^{n} (v_{i,j} \cdot \alpha_{i,j})) \cdot P_{pub} + \sum_{j=1}^{n} (v_{i,j} \cdot \beta_{i,j} \cdot K_{i,j})$$

From the derivation, if formula (1) is true, it means that the digital signatures in the data packages are verified successfully, that is, all data packets received by the fog node are valid.

6.2 Efficiency Analysis

In the proposed scheme, we will analyse the efficiency from the aspects of computation cost and communication cost. For the computation cost, we mainly analyse the time cost of vehicle in vehicle data collection phase and fog node training phase. We construct non-singular elliptic curve $y^2 = x^3 + ax + b \bmod p$ of order q in experiment, where p and q are prime numbers with a length of 160 bits, $a, b \in \mathbb{Z}_p^*$. Under the experimental environment of win 10 operating system, Intel Core i5-4210 processor and 2.8 GHz main frequency, the average value is obtained through 500 experiments in MIRACL. The average execution time of each cryptographic operation is shown in Table 1. Since the addition and multiplication of integers are less than other operations, we ignored them in the following analysis.

Table 1. Average execution time of cryptographic operations

Symbol	Meaning	Execution time/ms
T_{sm}	Scale multiplication	0.4420
T_{pa}	Point addition	0.0018
T_{ssm}	Small scale multiplication	$0.0138(t = 5)$
T_h	General hash operation	0.0001

In the stage of vehicle data collection, there are two parts of pseudonym generation and data package generation. The former needs two scale multiplication operations on the elliptic curve and two hash operations, while the latter needs one scale multiplication on the elliptic curve and hash operation. Therefore, the computation cost in this stage is $2T_{sm}+2T_h+T_{sm}+T_h = 3T_{sm}+3T_h = 1.3263$ ms.

In the phase of fog node training, the fog node batch verification needs $(n + 2)$ scale multiplication operations on the elliptic curve, $2n$ small scale multiplication operations on the elliptic curve, $(3n - 1)$ point addition operations on the elliptic curve and $2n$ hash operations. The computation cost is $(n+2)T_{sm} + (2n)T_{ssm} + (3n - 1)T_{pa} + (2n)T_h = 0.4752n + 0.8822$ ms. With the increase of the number n of batch verification data packages, the batch verification time of the scheme also increases. Figure 5 shows the comparison of the batch verification time of the scheme and other schemes with the increase of n.

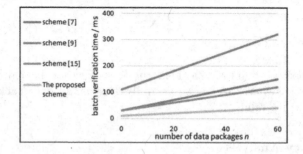

Fig. 5. Batch verification time of multiple data packages

As can be seen from the above figure, with the increase of the number of data packages n, the total time of batch verification of 60 data packages in the proposed scheme is about 50 ms, which is lower than other conditional privacy preserving schemes. Therefore, our proposed scheme can meet the computing needs in Vehicle Networks environment.

For the communication costs, on the vehicle side, the vehicle will generate data packages $DP_{i,j} = \{M_{i,j}, AID_{i,j}, T_{i,j}, K_{i,j}, \sigma_{i,j}\}$ after collecting traffic data. We assume that the traffic information transmitted by each scheme is the same, so $M_{i,j}$ do not make analysis and comparison. $AID_{i,j} = (AID_{i,j,1}, AID_{i,j,2})$, $AID_{i,j,1}, AID_{i,j,2}, K_{i,j} \in \mathbb{G}$ and $\sigma_{i,j} \in \mathbb{Z}_q$, where the elements in $\mathbb{G}$ are 40 bytes, let the output of hash function and time-stamp be 16 bytes and 4 bytes. Therefore, the communication overhead of the above data package is 40 * 3 + 20 + 4 = 140 bytes. Table 2 shows the comparison of communication cost between the proposed scheme and other schemes.

Table 2. comparison of communication cost

	Sending one data package/byte	Sending n data packages/byte
Scheme [7]	388	$388n$
Scheme [9]	644	$644n$
Scheme [15]	388	$388n$
The proposed scheme	140	$140n$

It can be seen from Table 2 that the communication overhead of this scheme is far less than that of [1], and less than that of [20] and [23]. When the number of data packages n increases, the corresponding communication overhead will also increase. Figure 6 is a comparison diagram of the communication overhead of each scheme with the increase of the number of data packages n. Therefore, the proposed scheme is more efficient in Vehicle Networks.

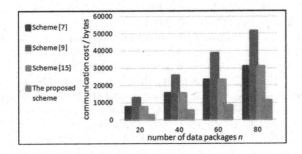

Fig. 6. Comparison of the communication cost

7   Conclusion

This paper proposes a conditional privacy preserving scheme which integrates federated learning into Vehicle Networks. This scheme can make the processing and analysis of Vehicle Networks models more efficient and secure. The identity of the vehicle is based on the pseudonym mechanism, which enables the vehicle to transmit messages safely. The training of the model is done in a distributed manner, which is faster and more efficient. The outsourced cloud server is used to aggregate all local models to reduce the workload of trusted authority. After obtaining the global model, the trusted authority can carry out macro-control and achieve flexible and targeted control. The security analysis shows that the scheme meets the conditional privacy preserving and other desirable security properties, which can not only prevent various attacks from external adversaries, but also resist internal adversaries such as semi-honest cloud server and colluding fog nodes. The efficiency analyses show that the proposed scheme reduces the computing costs and communication costs compared with the existing related schemes, improves the overall efficiency of system operation, and makes the proposed scheme more suitable for deployment in Vehicle Networks.

Acknowledgement. This work was partially supported by the National Natural Science Foundation of China (Grant No. 62072134, 61702168, U2001205), Key Research and Development Program of Hubei Province (Grant No. 2020AAA001), and Guangxi Key Laboratory of Trusted Software (Grant No. KX201908). We are also grateful to the anonymous reviewers for their valuable comments on the paper.

References

1. Bayat, M., Barmshoory, M., Rahimi, M.: A secure authentication scheme for VANETs with batch verification. Wireless Netw. **21**(5), 1733–1743 (2015)
2. Bo, X.-L.: Analysis of the security mechanism of internet of vehicles communication based on public key cryptography. Internet Things Technol. **5**(8), 32–34 (2018)
3. Bresson, E., Catalano, D., Pointcheval, D.: A simple public-key cryptosystem with a double trapdoor decryption mechanism and its applications. In: Laih, C.-S. (ed.) ASIACRYPT 2003. LNCS, vol. 2894, pp. 37–54. Springer, Heidelberg (2003). https://doi.org/10.1007/978-3-540-40061-5_3
4. Cao, B., Sun, Z.-H., Zhang, J.-T.: Resource allocation in 5G IoV architecture based on SDN and fog-cloud computing. IEEE Trans. Intell. Transp. Syst. **22**, 1–9 (2021)
5. Hao, F.: A 2-round anonymous veto protocol. In: Christianson, B., Crispo, B., Malcolm, J.A., Roe, M. (eds.) Security Protocols 2006. LNCS, vol. 5087, pp. 212–214. Springer, Heidelberg (2009). https://doi.org/10.1007/978-3-642-04904-0_29
6. Gentry, C.: Certificate-based encryption and the certificate revocation problem. In: Biham, E. (ed.) EUROCRYPT 2003. LNCS, vol. 2656, pp. 272–293. Springer, Heidelberg (2003). https://doi.org/10.1007/3-540-39200-9_17
7. Gu, C.-X., Zhu, Y.-F., Pan, X.-Y.: Forking lemma and the security proofs for a class of ID-based signatures. J. Softw. **18**(4), 1007–1014 (2007)
8. He, D.-B., Zeadally, S., Xu, B.-W.: An efficient identity-based conditional privacy-preserving authentication scheme for vehicular ad hoc networks. IEEE Trans. Inf. Forensics Secur. **10**(12), 2681–2691 (2015)
9. Horng, S.-J., Tzeng, S.-F., Huang, P.-H., Wang, X.-M.: An efficient certificateless aggregate signature with conditional privacy-preserving for vehicular sensor networks. Inf. Sci. Int. J. **317**(3), 48–66 (2015)
10. Hwang, J.-Y., Song, B., Choi, D.: Simplified small exponent test for batch verification. Theoret. Comput. Sci. **662**, 48–58 (2016)
11. Kamat, P., Baliga, A., Wade, T.: An identity-based security framework for VANETs. In: International Workshop on Vehicular Ad Hoc Networks. ACM (2006)
12. Li, J.-L., Choo, K.-K.-R., Zhang, W.-G.: EPA-CPPA: an efficient, provably-secure and anonymous conditional privacy-preserving authentication scheme for vehicular ad hoc networks. Veh. Commun. **13**(1), 104–113 (2018)
13. Liu, H., Li, H.: A distributed authentication protocol for VANET. J. Xi'an Jiaotong Univ. **47**(2), 58–62 (2013)
14. Lu, R..-X.., Lin, X.-D., Luan, T.-H.: Pseudonym changing at social spots: an effective strategy for location privacy in VANETs. IEEE Trans. Veh. Technol. **61**(1), 86–96 (2012)

15. Lu, R.-X., Lin, X.-D., Zhu, H.-J.: ECPP: efficient conditional privacy preservation protocol for secure vehicular communications. In: INFOCOM the Conference on Computer Communications IEEE. IEEE (2008)
16. Ng, J.-S., Lim, W.-Y.-B., Dai, H.-N.: Communication-efficient federated learning in UAV-enabled IOV: a joint auction-coalition approach. In: GLOBECOM 2020–2020 IEEE Global Communications Conference. IEEE (2020)
17. Palanisamy, B., Liu, L.: MobiMix: protecting location privacy with mix-zones over road networks. In: Proceedings of the 27th International Conference on Data Engineering, ICDE 2011, Hannover, Germany, 11–16 April 2011. IEEE (2011)
18. Raya, M., Jean-Pierre, H.: Securing vehicular ad hoc networks. J. Comput. Secur. **15**(1), 39–68 (2007)
19. Shao, J., Lin, X.-D., Lu, R.-X.: A threshold anonymous authentication protocol for VANETs. IEEE Trans. Veh. Technol. **65**(3), 1711–1720 (2015)
20. Shim, K.-A.: CAPS: an efficient conditional privacy-preserving authentication scheme for vehicular sensor networks. IEEE Trans. Veh. Technol. **61**(4), 1874–1883 (2012)
21. Wang, M., Wu, J., Li, G.-L.: Toward mobility support for information-centric IoV in smart city using fog computing. In: 2017 IEEE International Conference on Smart Energy Grid Engineering (SEGE). IEEE (2017)
22. Xu, G.-W., Li, H.-W., Liu, S.: VerifyNet: secure and verifiable federated learning. IEEE Trans. Inf. Forensics Secur. **15**, 911–926 (2019)
23. Rong, Yu., Kang, J.-W., Huang, X.-M.: MixGroup: accumulative pseudonym exchanging for location privacy enhancement in vehicular social networks. IEEE Trans. Dependable Secure Comput. **13**(1), 93–105 (2016)
24. Zhang, J., Cui, J., Zhong, H.: PA-CRT: Chinese remainder theorem based conditional privacy-preserving authentication scheme in vehicular ad-hoc networks. IEEE Trans. Dependable Secure Comput. **18**(2), 722–735 (2019)
25. Zhang, Z.-H., Li, Q.-D., Fu, Y.: Adaptive federated deep learning with non-IID data. Acta Automatica Sinica, 1–13 (2021)
26. Zhou, C.-Y., Fu, A.-M., Shui, Yu.: Privacy-preserving federated learning in fog computing. IEEE Internet Things J. **7**(11), 10782–10793 (2020)
27. Zhu, C.-S., Liu, P.-H., Wang, Q.-R.: Group key management scheme based on group signature with fault tolerance for mobile ad hoc networks. Appl. Res. Comput. **28**(10), 3811–3816 (2011)

Dissecting Membership Inference Risk in Machine Learning

Navoda Senavirathne[1]([envelope]) [iD] and Vicenç Torra[2] [iD]

[1] School of Informatics, University of Skövde, Skövde, Sweden
navoda.senavirathne@his.se
[2] Department of Computer Science, University of Umeå, Umeå, Sweden
vicenc.torra@umu.se

Abstract. Membership inference attacks (MIA) have been identified as a distinct threat to privacy when sensitive personal data are used to train the machine learning (ML) models. This work is aimed at deepening our understanding with respect to the existing black-box MIAs while introducing a new label only MIA model. The proposed MIA model can successfully exploit the well generalized models challenging the conventional wisdom that states generalized models are immune to membership inference. Through systematic experimentation, we show that the proposed MIA model can outperform the existing attack models while being more resilient towards manipulations to the membership inference results caused by the selection of membership validation data.

Keywords: Data privacy · Privacy preserving machine learning · Membership inference attack

1 Introduction

Privacy in ML has become a forefront topics in recent years with the increased use of sensitive, personal data to train the ML models. ML models, especially deep neural networks (DNN) are known to *memorize* their training data when they have a sufficient capacity (i.e., number of trainable parameters in DNNs are higher than the number of training data instances) [19]. This memorization capability is exploited by several types of attack models to infer information about the underlying training data leading to privacy vulnerabilities. A growing body of prior work have shown that ML models leak private information about the underlying training data though their observable outputs and optimization information [3,12,14,15,17,18]. *Membership inference* is one of the most discussed privacy vulnerabilities in ML [12,14,15,18]. In this case, the adversarial objective is to distinguish whether a given record is part of the ML model's training set or not. For instance, when a target ML model is built on healthcare information, by simply revealing membership the specific health conditions of the individuals whose data are used to train the ML model can be disclosed leading to a severe violation of their privacy. Nevertheless, it is shown that membership

© Springer Nature Switzerland AG 2022
W. Meng and M. Conti (Eds.): CSS 2021, LNCS 13172, pp. 36–54, 2022.
https://doi.org/10.1007/978-3-030-94029-4_3

inference can be successfully carried out as a black-box attack with a minimal amount of information which highlights the severity of this risk even further.

In many of the black-box membership inference methods, the adversaries exploit the prediction confidence score vectors/soft-max probabilities produced by the target ML model with respect to a given feature vector to determine its membership. Numerous, studies have proposed variants of this approach to realize membership inference. The first black-box MIA was proposed by Shokri et al. [15] made use of a neural network (NN) based binary classifier to determine membership. However, as shown by Song et al. [16], the use of customized NN based classifiers underestimate the membership risk due to misappropriated hyper-parameter settings such as the number of hidden layers and learning rate. Therefore, non DNN based MIAs are proposed in the literature which are based on a threshold value in order to distinguish between members and non-members [6,16,18].

Many studies have identified that *model over-fitting* as the most common cause for membership inference [8,15]. A ML model is said to over-fit to its training data when its performance on unseen test data is significantly poor compared to training data. This is also known as the generalization gap. A myriad of mitigation techniques is proposed for back-box MIAs. They are leveraging different approaches to maintain indistinguishably between members and non-members thus increasing the uncertainty for the membership inference adversary. Differential privacy is applied to ML at the training time to limit the effect of a single data instance has on the overall model output. It is shown that differential privacy can be used as a mitigation technique against MIAs as reducing the effect of a single data instance on the model output make inferring the presence of that specific instance more challenging [13]. However, Humphries et al. [5] suggest that certain properties of the underlying data set, such as bias or data correlation, plays a critical role in determining whether differential privacy is effective as a mitigation strategy for MIAs. Nasr et al. [11] propose a method to train the target ML model to simultaneously achieve two objectives as, maintaining model's prediction accuracy while lowering the MIA accuracy. This is done via adding the inference attack as an adversarial regularization term. Another mitigation strategy is proposed by Jia et al. [7] known as *MemGuard*. In this case, the model's prediction outputs are obfuscated with noisy perturbations such that members and non-members are indistinguishable based on the perturbed outputs. However, some recent work has shown that both of above mentioned mitigation strategies are not effective with respect to novel MIAs [2,16]. The inconsistent performance of the existing attack models and the ineffectiveness of some of the proposed mitigation strategies for MIA show that we still have not properly understood membership inference risk.

Hence, in this work, we aim at deepening our understanding of the existing black-box MIA models. Our main contributions are briefly explained below. First, we present a brief survey on the existing black-box MIAs and experimentally evaluate them for their effectiveness. Then, we propose a new MIA model that outperforms the existing black-box MIA models with respect to both generalized and over-fitted ML models. Then, confirming the findings of related works [8,18] we find that MIAs based on classification correctness perform comparably with the rest of the more sophisticated attack models. Hence, we systematically investigate the relationship between classification correctness and the membership inference which is not attempted by the previous work. Based on the aforementioned findings we showcase how the MIA's results can be manipulated by the selection of the data included in the membership validation set that may lead to overstating or understating the membership risk.

2   Related Work

Membership inference attacks (MIA) aim at identifying whether a given data instance is included in the training set of a target ML model or not. Membership identification can bring a distinctive risk to privacy as it may reveal sensitive information about the individuals whose data are used to train the ML model. MIAs are shown to be very effective as they are applicable to different ML algorithms (e.g., deep neural networks, decision trees, logistic regression etc.), different model architectures, different model types (e.g., classifiers, generative models, and sequence to-sequence models), different ML setups (e.g., centralized ML, federated ML), and with different levels of adversarial knowledge (e.g., black-box, white-box). Out of these, the main categorization for MIAs is based on adversarial knowledge. In this Section, we present an overview of the existing MIAs based on adversarial knowledge. Most of the existing MIAs assume that the adversary has the knowledge of training data distribution, the knowledge of the training data process, the knowledge of the model architecture, and the knowledge of the target ML model' s output. Whereas, based on the availability of the knowledge of model parameters MIAs can be categorized as *black-box* or *white-box* attacks.

In black-box MIAs, it is assumed that the adversary has query access to the target ML model f. That is, for a given input instance z the adversary can obtain the output of the target ML model which could be either full, partial or label only output. The basic argument behind the realization of MIAs is that the target ML models behave differently on the data they have been exposed to at the training time (i.e., members/training data) as opposed to the data they have not been exposed to (i.e., non-members/test data). It is shown that the differences in the target ML model's behaviour is reflected via the model's output and can be leveraged to distinguish between members and non-members. In this paper, we discuss about the black-box MIAs targeting DNNs. We believe this choice is appropriate as DNNs are increasingly used to solve complex real world problems in privacy sensitive nature, and the black-box MIAs are identified as a severe threat to the privacy of personal data.

2.1   Customized Neural Network Based Classifiers

In this case, a DNN based binary classifier is trained as the membership attack model which takes the prediction confidence score vector obtained from the target ML model f for a given input to determine its membership. To train the attack model for membership inference, the shadow training approach is proposed by Shokri et al. [15]. First, the adversary trains multiple shadow models to mimic the target model's behaviour. These shadow models are trained on shadow datasets (S) drawn from the same distribution as the target ML model's training data Ψ. To make the attack non-trivial it is assumed that the shadow data and target model's training data are disjoint or it can be denoted that $S \sim \Psi \cap D \sim \Psi = \emptyset$ (nevertheless, Shokri et al. propose a mechanism to generate synthetic data in case the adversary does not have access to any shadow data). Once the shadow models are trained, prediction output vectors are obtained using each of the shadow models on their own training dataset and a separate test dataset. Then the prediction output vectors with respect to the shadow training set are labeled as "members" whereas outputs of the shadow test set are labeled as "non-members" for each shadow model respectively. These data constitute the training set for the NN based binary classifier or the membership inference attack model. Further, multiple attack models are generated to infer the membership as one per each class. To execute the MIA, the adversary queries the target ML model with a data record and obtains the prediction confidence score vector. Then it is passed to the attack model along with the true label to infer the membership status.

2.2   Confidence Score Based Membership Inference

In these attack models, the membership determination is done based on a threshold value computed over the confidence score vector output by the target ML model. There exist multiple metrics that can be derived using the confidence score vector and the ground truth label of a given instance such as maximum confidence score, entropy or loss. As explained in Sect. 1 the use of customized NN based classifiers for membership inference is said to underestimate the membership risk due to misappropriate hyper-parameter settings [16]. Moreover, Salem et al. [14] argue that requirements like number of required shadow models used in the aforementioned attack model are too strong for such an attack model to be realistic. Therefore, confidence score threshold based attacks are more preferred due to their simplicity and consistency of the results.

Prediction Confidence: Use of prediction confidence score values for determining the membership is based on the intuition that the target ML model is trained by minimizing prediction loss over training data. Thus, the prediction confidence of a data instance included in the training set should be higher than the prediction confidence with respect to a data instance not included in the training set. Based on the above assumption, the adversary infers an input example as a member if its prediction confidence is higher than a given threshold. This attack model is shown to be reasonably effective by Yeom et al. [18], where the threshold values are learned through shadow model training. Yeom et al. use a single threshold for all the class labels whereas Song et al. [16] propose an improved method where different threshold values are used for different class labels.

If $\mathbf{1}\{.\}$ is the indicator function and τ_y represents the class based threshold value, then the attack model can be summarized as, $I_{conf}(f,(x,y)) = \mathbf{1}\{f(x)_y \geq \tau_y\}$.

Prediction Entropy: Typically, the target ML models are trained by minimizing the prediction loss over training data. This indicates that the prediction confidence of a given training instance should be close to the one-hot encoded vector and its prediction entropy should be close to 0. Whereas, for the unseen samples, the target ML model usually has a larger prediction entropy. As proposed by Song et al. [16], we can rely on prediction entropy for membership inference. The adversary determines an input example as a member if its prediction entropy is smaller than a given threshold, a non-member otherwise. Shadow training technique can be used for computing the class wise threshold entropy values. The attack model can be summarized as, $I_{entr}(f,(x,y)) = \mathbf{1}\{-\sum_i f(x)_i \, log(f(x)_i) \leq \tau_y\}$ where $\mathbf{1}\{.\}$ is the indicator function and τ_y represents the class based entropy threshold value.

Further Song et al. point out a major limitation of entropy based MIA. That is, it does not take the true label values into consideration. Therefore, they present a *modified entropy* based MIA in [16] where an accurate classification with probability of 1 leads to modified entropy 0, while an incorrect classification with probability of 1 leads to modified entropy infinity. The attack model can be summarized as below.

$$I_{Ment}(f,(x,y)) = \mathbf{1}\{Mentr\,(f(x),y) \leq \tau_y\}. \tag{1}$$

Here, $Ment\,(f(x),y)$ is computed as below.

$$I_{Ment}(f(x),y)) = -(1 - f(x)_y)\, log(f(x)y) \; - \sum_{i \neq y} f(x)_i \, log(1 - f(x)_i). \tag{2}$$

Prediction Loss: Yeom et al. [18] presented a simple, less computationally intensive approach to determine membership based on per instance loss. A data instance is identified as a member if its prediction loss is smaller than the expected training loss (average loss of the target ML model). At the training phase, the ML model is trained by minimizing the prediction loss over the training data. Thus it is expected that the prediction loss of a training sample is smaller than that of a test sample (a sample not used for training). The attack model can be summarized as, $I_{Loss}(f(x),y)) = \mathbf{1}\{\mathcal{L}(f(x),y) \leq \tau_y\}$ where $\mathbf{1}\{.\}$ is the indicator function and τ_y represents the class wise loss threshold. However, the assumption of having access to the average training loss of a target ML model is controversial. Therefore, the use of shadow training can be used to obtain an approximation of the loss threshold.

2.3   Label Only Membership Inference

Recent work has showcased that some of the label only attacks perform in par with the above discussed confidence score based attack models, whereas others surpass them. In other words, it is shown that the predicted labels are effective enough to reveal membership. Below we describe some of the label only attacks briefly.

Classification/Prediction Correctness: This attack depends on a naive approach to predict membership. Here, correctly classified data instances are categorized as members while incorrectly classified instances are classified as non-members. Use of prediction correctness to determine membership has shown to achieve comparable success as the use of customized NN based attack [16] and confidence score based attacks [2,9]. The intuition here is that the target ML model has memorized its training data as a result of over-fitting thus failing to generalize well on the test data. In such a scenario, the chances are high that the prediction accuracy reveals true members. Where $\mathbf{1}\{.\}$ is the indicator function, the attack model can be summarized as $corr(f, (x, y)) = \mathbf{1}\{argmax_i \ f(x)_i = y\}$.

Boundary Distance Based: These attacks are aimed to predict the membership using a given data instance's distance to the target model's decision boundary. Both Choquette-Choo et al. [2] and Li et al. [9] propose this attack model based on the intuition that the over-fitted ML models make more confident decisions with respect to their training data as opposed to test/non-member data. Assume a training data instance x belongs to class y and the target model correctly classifies it too. Then we perturb x as $\hat{x} = x + \epsilon$, so that the target ML model would now predict a different label for $\hat{x}$ as $f(\hat{x}) \neq y$. Assuming the target model is more confident about its decisions with respect to the training data, the amount of perturbation required to change its decision (i.e., predicted label) is higher than that of a non-member instance.

If we obtain an estimated $L2-$distance to the decision boundary of the target model with respect to a data instance x as $d(x, y)$, then x will be classified as a member, if $d(x, y) > \tau$. For tuning the threshold τ, Choquette-Choo et al. [2] propose shadow training. Whereas, while Li et al. [9] relax the shadow data requirement and propose a method to use random data points instead. Different methods are proposed in the above two works to generate perturbed data in order to determine the decision boundary such as data augmentation, random noise and decision boundary based method. Here, we explain the decision boundary based method as the other two methods are intuitive.

The decision boundary based method is a black-box mechanism for generating adversarial examples with minimal adversarial knowledge. In this case, it is assumed that the adversary only has access to the target model's predictions. Both [2,9] make use of a variant of this attack known as *HopSkipJump* attack introduced in [1]. To initiate the HopSkipJump attack, a random point x' is picked with respect to a given data instance (x, y) such that $f(x') \neq y$. Here, x' is considered as the initial adversarial example. Then the algorithm performs a random *walk* along the boundary between the adversarial and its non-adversarial region such that the distance towards x is minimized with each iteration while still remains on the adversarial region. By iteratively querying the target model the adversary can obtain the decision boundary with respect to a given data instance in terms of its nearest adversarial examples. The idea leveraged by all of the above mentioned attacks is that the training data are more robust towards perturbations (i.e., augmentation, random noise, adversarial examples etc.). Thus, it requires more effort to modify them in a way to produce samples that would get incorrectly classified by the target model compared to test data. As explained by [2], computing the distance from a given data instance to the decision boundary is exactly the problem of finding the smallest adversarial perturbation. Hence, the data instances with a higher distance to the decision boundary (i.e., nearest adversarial instances) are classified as members. All the above discussed attacks required multiple queries per each data instance sent to the target model in order to obtain their predictions. More specifically, it is shown that the decision boundary based attack requires about $\approx$2000 queries per data instance in order to achieve comparable accuracy with customized NN based MIA model.

Transfer Attack: This attack model is proposed by Li et al. [9]. In this case, it is assumed that the adversary has access to a shadow dataset S that comes from the same distribution as the target model's training data. First, the adversary queries the target model with the shadow data and get them relabeled. Then the relabeled data are used to train a shadow model to mimic the target model's behaviour. The intuition here is that by relabeling the shadow data the adversary can obtain a model much closer to the target model which may also contain sufficient membership information. Then the correctly classified data are identified as members and others as non-members.

3 Notations

Let f be a DNN classifier trained on dataset D by minimizing the average prediction loss over all the training samples as $min \ \frac{1}{|D|} \sum_{z \in D} \mathcal{L}(f, z)$. Here, $|D|$ denotes the size of D and $\mathcal{L}(.)$ denotes the loss function. Data instance $z \in D$ can be expressed as $z = (x, y)$ where x and y respectively indicates the feature vector and the target/class variable. Here, $z \sim D$ and the training dataset (members) can be denoted as $D \subseteq \Psi$. The test dataset (non-members) can be denoted as $D' \subseteq \Psi$ where $D \cap D' = \emptyset$. ML model f learned over D outputs a prediction confidence vector over all the k class labels as $\sum_{i=1}^{k} f(x)_i = 1$. Then the class label associated with the highest classification confidence is output as the classification result for x which can be indicated as $\hat{y} = argmax \ f(x)_i$ where $\hat{y}$ is the predicted class. A soft-max function is often used to map a ML model's outputs to a probability distribution or the classification confidence score vectors. For a given a model f soft-max values are calculated as, softmax $f(u)_i = \frac{e^{u_i}}{\sum_{j=1}^{k} e^{u_i}}$. Here, u_i denotes the input vector passed to the soft-max function which is in fact the output of the penultimate layer of f concerning a particular input x_i. Then the standard exponential function is applied to each element of the vector u_i. The denominator indicates the normalization term which ensures that all the output values of the soft-max function will sum up to 1 thus producing a valid probability distribution.

4   Activation Pattern Based MIA

As explained earlier the existing MIA models are shown to be more successful with respect to over-fitted ML models and they perform poorly with respect to well generalized ML models. Thus, in the real world scenarios where generalized ML models are deployed the usefulness of these attack models become questionable. However, this does not necessarily mean that well generalized models are completely immune to MIAs. Long et al. [10] have shown that with respect to the input data instances with a unique influence on the target ML model (i.e., outliers), the well generalized models could also be vulnerable to MIAs.

In this Section, we introduce a novel MIA based on the concepts of transfer attack and model activations with the aim of improving the performance of membership inference on well generalized ML models. In this case, it is assumed that the adversary only has access to the ground truth labels (not the confidence scores), the target model's architecture and the shadow data drawn from the same distribution as the target ML model without any over-lapping. We explain the different steps with respect to the MIA model under the below headings.

- **Shadow data relabelling and shadow model training -** Following the same procedure as transfer attack the adversary gets the shadow data (S) relabeled using the target model f as $\hat{S} = f(S)$. Then a part of relabelled data $\hat{S}_{tr}$ is used to train a shadow model locally at the adversary's site as $\hat{f}_s$. Due to shadow data relabelling, we assume that the shadow model represents the target ML model more closely specifically in the case of well generalized models.
- **Extracting penultimate layer activation -** The proposed MIA model is based on exploiting the penultimate layer outputs extracted from $\hat{f}_s$ with respect to a given dataset. This approach is shown to be very effective as a white-box MIA when the adversaries have access to the target model's penultimate layer outputs [12]. As explained by Nasr et al. [12] the activation functions in the penultimate layer of a DNN extract more complex features with respect to its training data thus leaking more membership information. Therefore, leveraging the penultimate layer's output to determine membership seems sensible. Due to relabelling of shadow data the shadow model $\hat{f}_s$ works as a surrogate for target model f. Therefore, it is intuitive that the penultimate layer's outputs between the two models are similar with respect to identical input data. Based on this intuition we devise a method to distinguish between members and non-members in the next step.
- **Activation pattern based anomaly detection -** To distinguish between members and non-members we use a simple distance based anomaly detection technique which considers the distance of the kth nearest neighbor of a given data instance x as an outlier score w [20]. The outliers are those points that have large values for w. In this case, we utilize the activation patterns extracted from the shadow model $\hat{f}_s$ with respect to relabelled shadow data $(\hat{S})$ by making a forward pass through the trained DNN $\hat{f}_s$. Then the extracted activation patterns are used to train class wise k-nearest neighbour

classifiers (KNN) for outlier detection. In other words, for each class label c, a KNN model is generated with a user specified number of neighbours k. Next, the activations are extracted from $\hat{f}_s$ with respect to the membership validation set V which contains both members and non-members. The adversary's objective here is to identify true members and non-members from V. In the next step, trained KNN classifiers for a specific classes (KNN_c) are used to classify the instances in V that belongs to the same class as KNN_c as *outliers* or *inliers*. Here, outliers are identified as non-members and inliers as members.

- **Improving membership inference accuracy by noise addition -** By adding carefully calibrated noise to membership validation dataset V we show that the accuracy of the membership inference can be further improved with respect to well-generalized models. The well-generalized models perform well with respect to both member and non-member data. This is due to the fact that they have learned more generalizable patterns from the underlying training/member data. Thus differentiation between members and non-members becomes challenging. Therefore, with respect to well-generalized target models, we add a carefully calibrated noise to V. The intuition here is that by adding noise to member data the deviation that occurs to the model's activation patterns are limited compared to adding noise to the non-member data. Hence the distance to the kth neighbour would be higher for non-members compared to members making it easier to distinguish between members and non-members. However, with respect to over-fitted models, such noise addition is not necessary as each data instance is more likely to have a unique activation pattern (due to the over-fitted ML models tend to memorized information about the input data instead of learning a generalized pattern).

- **Calibrating noise and neighbourhood size -** By following the shadow training procedure we decide the best noise level α and the neighbourhood size k. For well generalized models a neighbourhood size of 5 seems to be appropriate whereas for over-fitted models the neighbourhood size of 1 seems to be appropriate. For image and Adult data a random Gaussian noise is added with $\mu = 0$ and standard deviation σ as $x + \mathcal{N}(\mu, \sigma^2)$. And the standard deviation (σ) is chosen based on the shadow training procedure. For binary datasets, we instead compute a Bernoulli noise where the attribute values are flipped with probability p. p is also chosen based on shadow training. In each of these cases, the σ and p values are chosen to maximize both precision and the recall values.

5   Experimental Setup

This Section explains the experimental setup. Table 1 contains a brief description of the classification datasets used for empirical evaluation.

Table 1. Dataset description. $|D|$ and $|T|$ respectively denote the size of the training and test datasets whereas $|S|$ indicates the size of the shadow dataset.

Dataset	Data type	& Instances × &attributes	& Classes	$|D|$ and $|T|$	$|S|$
MNIST	28 × 28 images	70, 000	10	20, 000	10, 000
FMNIST	28 × 28 images	70, 000	10	20, 000	10, 000
Adult	tabular+mix	48, 842 × 14	2	10, 000	5, 000
Purchase-100	tabular+binary	197, 324 × 600	100	20, 000	10, 000
Texas-100	tabular+binary	67, 330 × 6, 170	100	10, 000	10, 000
Location-30	tabular+binary	5, 010 × 446	30	1, 200	600

For Texas-100 and Purchase-100 datasets, we use the same ML model architecture as mentioned by Nasr et al. [12]. For Purchase-100 dataset we use a fully connected DNN model with layer sizes of 600, 1024, 512, 256, 128, 100 (where 100 is the output layer). For Texas-100, we used layers with size 1024, 512, 256, 128, 100 (where 100 is the output layer). Adam optimizer is used for learning the models with 100 epochs. For MNIST and FMNIST datasets we used a convolutional neural network (CNN) with 2 layers of size 32 and trained for 50 epochs with SGD optimizer with learning rate 0.01. Also, we used $l2$ regularization with a weight of 0.001. We use a balanced training dataset with respect to all the datasets except for the purchase-100 dataset. To evaluate the attack models we employ accuracy, precision and recall.

6   Experimentation

6.1   Evaluating MIAs

In this sub-section, we first evaluate the accuracy of the aforementioned black-box MIAs. For measuring the effectiveness of MIAs we use a balanced membership validation dataset V with 50% member and 50% non-members. With respect to a balanced membership validation dataset, a membership inference accuracy closer to 0.5 would indicate that distinguishing a member from a non-member is only as good as a random guess. Table 2 depicts the results of the different MIAs along with model accuracies. The results with respect to the boundary distance based (BDB) attack are extracted from [2] as the attack model is very expensive to execute. However, note that in this case the model's architectures and training/shadow data sizes are not the same as the rest of the MIA models. Nevertheless, we can use the results of the BDB attack model to draw a rough comparison.

With respect to the ACT attack model, we have to first distinguish between well-generalized models and over-fitted models. As explained in Sect. 4, in order to improve the membership inference accuracy with respect to well-generalized models we calibrate noise to each data instance in V. The size of the neighbour-hood is set to 5. Whereas, for over-fitted models, no noise is calibrated to V whereas the neighbourhood size is set to 1. One other important fact in this case is to identify whether a given model is over-fitted or not. Depending on generalization error (GE) for this might not be the best indication of the model's generalizability unless the target model's training accuracy is closer to 1. Hence, with respect to these experiments, we adopt a simple heuristic to decide the generalizability of the models. A modified generalization gap (MGE) is computed as $MGE = 1 - test\ error$. If $MGE \lesssim 0.1$ we consider that model as a generalized model and otherwise as an over-fitted model. However, in a real world scenario, an adversary might not have access to the test accuracy of the target model f. Hence, the test accuracy with respect to shadow model $\hat{f}_s$ can be used to decide whether a given ML model is generalized or not.

The membership inference accuracies averaged over all the aforementioned attack models vary as $0.533, 0.587, 0.565, 0.766, 0.667$, and 0.648 respectively for MNIST, FMNIST, Adult, Location-30, Texas-100 and Purchase-100 datasets. As observed by the previous works, a low generalization error leads to low membership inference accuracy. Confirming the same behaviour we can note that the models trained on MNIST, FMNIST and Adult datasets that have reported low generalization errors as $0.02, 0.1$ and 0.06 have also reported a low membership inference accuracy. The accuracies of the different MIA models averaged over the aforementioned datasets vary as $0.63, 0.61, 0.629, 0.593, 0.535, 0.604, 0.509, 0.789$, and 0.825 respectively for CONF, ENT, M-ENT, CuNN, LOSS, CC, TRANS, BDB, and ACT attack models. Among the different attack models, we can see that the ACT attack model outperforms the rest of the attack models in many cases (except for purchase-100). The main advantage of the ACT attack model over the BDB model is that we do not have to overburden the target ML model with repetitive queries. Instead, after the data relabelling, we can completely run the attack model locally. Whereas, compared to the rest of the attack models (i.e., CONF, ENT, M-ENT, CuNN, LOSS, CC, TRANS) the ACT attack model performs significantly better even with less amount of background knowledge (i.e., no confidence scores are required for the ACT attack). More importantly it can be noticed that the ACT model performs comparably with respect to both well generalized and over-fitted models. Li et al. [9] has shown that the TRANS attack model is very effective with respect to ML models with low generalization error. It is shown in with respect to image detests such as CIFAR-10, CIFAR-100, GTSRB, and Face the attack model achieves good accuracy. However, with respect to the datasets we experimented no significant results were observed. However, using TRANS attack model along with activation pattern based outlier detection (i.e., ACT model) has significantly improved the performance of membership inference.

Table 2. Summary of the target ML model accuracy (train and test) and MIA accuracy scores for each membership inference attack type, maximum confidence score (CONF), entropy based (ENT), modified entropy based (M-ENT), customized NN based (CuNN), loss based (LOSS), correct classification based (CC), transfer attack (TRANS), boundary distance based (BDB), and activation based (ACT).

DataSet	TRAIN	TEST	CONF	ENT	M-ENT	CuNN	LOSS	CC	TRANS	BDB	ACT
MNIST	0.997	0.977	0.51	0.506	0.511	0.512	0.502	0.505	0.496	0.578	**0.727**
FMNIST	1	0.893	0.584	0.566	0.585	0.574	0.5	0.548	0.496	NA	**0.839**
Adult	0.858	0.795	0.565	0.536	0.565	0.561	0.5	0.54	0.507	0.587	**0.723**
Location-30	0.995	0.505	0.811	0.781	0.805	0.739	0.602	0.749	0.523	0.892	**0.994**
Texas-100	0.806	0.437	0.668	0.647	0.667	0.563	0.57	0.671	0.531	0.803	**0.884**
Purchase-100	0.974	0.754	0.647	0.629	0.645	0.607	0.535	0.609	0.503	**0.874**	0.783

Then we have a closer look at the performance of the ACT attack model as summarized in Table 3. Here, we evaluate the attack model using the precision and recall metrics. Precision is the fraction of the records identified as members that are true members. Recall is the fraction of true members identified. Here, we can see the new attack model has high precision ranging from 0.68 to 1 with respect to all the datasets (including a precision score of 1 concerning the highly generalized two models, MNIST and FMNIST). The recall values for the highly generalized two models (i.e., MNIST and FMNIST) are lower than the rest. However, having a high precision value is more important in the case of membership inference as it indicates that the chances of a member identified by the attack model being a true member is high.

Table 3. Evaluating activation pattern based MIA (ACT) for precision and recall.

Dataset	Precision	Recall
MNIST	1	0.518 ± 0.08
FMNIST	1	0.694 ± 0.2
Adult	0.685 ± 0.07	1
Location-30	0.874 ± 0.15	1
Texas-100	0.695 ± 0.01	1
Purchase-100	0.791 ± 0.02	1

6.2 Classification Correctness and Confidence Scores

According to Table 2 it can be noticed that CC based MIA performs comparably with the rest of the more sophisticated attacks models based on the prediction confidence scores (i.e., CONF, ENT, M-ENT, CuNN, LOSS). The difference between best MIA accuracy and the CC based attack accuracy vary as $0.007, 0.037, 0.025, 0.062, 0$ and 0.038 respectively for MNIST, FMNIST, Adult,

Location-30, Texas-100, and Purchase-100 models. This shows that the advantage of using other attack models are relatively low (between $\approx 0.007 to 0.062$). CC attack's effectiveness in determining membership is directly linked to the generalization capability of the ML model. This means a model over-fitted to its training data is more likely to provide accurate classifications for the member/training data compared to the non-member/test data. Hence, model owners can obtain a fair understanding of the model's vulnerability towards membership inference by examining the generalization error of a given model if the training data used to build the model is representative enough (i.e., all subpopulations and classes are adequately represented in the training data). Compared to the aforementioned attack models CC based attack requires relatively little background knowledge (i.e., no requirement for data, training, and model knowledge. It only requires the ground truth label knowledge) and limited computations (i.e., no shadow training is required). Since CC based attack provides comparable results with the confidence score based methods we aim to explore the relationship between the prediction confidence scores and the classification correctness in this sub-section.

Typically, in data privacy literature it is understood that the over-fitted ML models *memorize* their training data and later this information are leaked through their observable outputs (i.e., classification confidence vectors). One essential requirement should be fulfilled in order for the classification confidence vectors to leak meaningful membership information. That is, the confidence assigned for member data should be *different* from that of the non-member data. The aforementioned MIAs such as CONF, ENT, M-ENT, and LOSS attacks assume that the members get classified with higher confidence compared to the non-members. Most of the works have visualized maximum confidence score distribution between members and non-members which seems to be different from each other. This justifies the use of confidence scores for membership inference. However, we argue that claiming confidence scores differ based on the membership of the data is more anecdotal than factual. Hendryck et al. [4] present an interesting view on the variability of the classification confidence scores. That is, the ML models tend to assign higher classification probabilities to correctly classified samples as opposed to the incorrectly classified samples. If this claim is true, it will explain, why having access to the classification confidence values do not necessarily improve the membership inference accuracy, and why CC based attack itself implies membership.

As explained earlier, in this work we consider the target ML model to be a DNN, which has a soft-max output layer. DNN's logits (u); output of the neurons in the penultimate layer; are fed into the soft-max function in order to obtain the classification confidence vector v as $v = \text{softmax}(u)$. Then for each prediction, we retrieve the maximum soft-max score (i.e., the soft-max score of the predicted class) output by the target ML model. Then we depict them using probability density plots respectively for each dataset. In each plot, we graph the classification confidence scores for members with correct classification (Members-CC), members with incorrect classification (Members-IC), non-members with correct classification (NonMembers-CC), and non-members with incorrect classification (NonMembers-IC). Further, on the same graph we plot the maximum soft-max scores for true members (true positives-TP) and true non-members (true negatives-TN) identified via M-ENT attack (we select this attack model as it turns out to be one of the best performing attack models based on the confidence score). These results are shown by Fig. 1. Here, it can be clearly observed that for the majority of the datasets the correctly classified instances are assigned high soft-max confidence values irrespective of their membership. In this case, the distribution of the confidence scores is similar to that of the identified true members. Whereas, when members and non-members get incorrectly classified, they show relatively low confidence which aligns with the distribution of the soft-max probabilities of the true non-members.

The only time a substantial difference between the member's confidence scores and the non-member's confidence scores are observed is with respect to the Location-30 dataset. The ML Model trained on the Location-30 dataset accounts for the highest generalization gap amongst all the datasets (i.e., 0.49). Even though the models trained on the Texas-100 and Purchase-100 datasets are also somewhat over-fitted (i.e., generalization gap greater than 0.2) we do not observe a significant difference in the soft-max confidence value distribution between members and non-members. This indicates that, unless a ML model is extremely over-fitted to its training data the confidence scores do not vary based on the membership of the data rather it is dependent on the classification correctness. However, in practice deploying such an extremely over-fitted model does not make any sense due to its poor generalizability. Hence, concerning the confidence score based MIA models the practicality of the membership inference risk seems a bit far fetched.

These results conclude that classification correctness is the dominating factor for the observable differences in the confidence scores, not their membership unless the target model is extremely over-fitted. Further, this supports our initial observation that why most of the black-box MIAs (i.e., CONF, ENT, M-ENT, LOSS) do not gain any significant advantage by having access to the classification probabilities for identifying true members. However, we cannot draw the same conclusion with confidence concerning the CuNN based attack model due to the fact that different attack model architectures (i.e., neural networks) could be able to pick non-trivial membership information. However, tuning such a model remains a question that needs addressing.

6.3   Manipulating Membership Inference Results

In the literature, results of different MIAs are used as a measure the privacy preservation with respect to a given ML model. The researchers use the performance of the MIAs either to evaluate the effectiveness of the proposed mitigation strategies against membership inference or to showcase the privacy risk of a given ML model. The most common evaluation measures of MIA are accuracy and precision. Under a balanced membership validation set (i.e., 50% members and 50% non-members) the use of MIA's accuracy is a common practice to evaluate the MIA model's effectiveness. In the previous sub-section, we showed

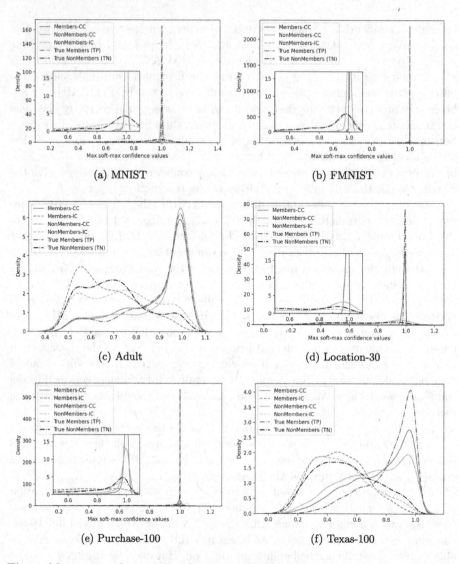

Fig. 1. Maximum soft-max classification probabilities for each dataset. (Here CC stands for correct classification and IC stands for incorrect classification.)

that a prominent relationship exists between the classification correctness and the confidence scores produced by a target ML model. A ML model trained by minimizing the average training loss over a well represented data sample will provide correct classifications to data instances similar to the underlying training dataset. Whereas, incorrect classifications can occur if the target ML model is not familiar with the characteristics of a given data instance due to lack of exposure to such data during training. Due to this, in membership inference, we come across false positives (FP-non-members in-correctly identified as members) and false negatives (members incorrectly identified as non-members). Generalizability of the target model is the main reason for FPs and under represented sub populations, outliers, or insufficient training can be considered as the main reasons for FNs.

Hence, it is intuitive that a target model can correctly classify any data instance that is similar to its average training data. Hence, we hypothesize that by carefully selecting member and non-member data for the membership validation set the membership privacy risk of the target ML model can either be understated or overstated. To illustrate this we set up the below experiments.

- Correctly classified non-members (CC-NM): In the first case, we cherry-pick the non-members for the membership validation set which are correctly classified by the target model.
- Incorrectly classified non-members (IC-NM): In the second case, we cherry-pick the non-members for the membership validation set which are incorrectly classified by the target model.

Table 4 showcase the MIA results with respect to the aforementioned experiment. At a glance, we can notice that by selecting correctly classified non-members (CC-NM) to the membership validation set the accuracy of the MIAs are understated compared to selecting incorrectly classified non-members (IC-NM). When membership validation dataset V is chosen randomly the average MIA accuracy vary as $0.533, 0.587, 0.56, 0.75, 0.65$, and 0.61 respectively for MNIST, FMNIST, Adult, Location-30, Texas-100 and Purchase-100 datasets (without BDB attack). Whereas, when the non-members included in V are all correctly classified instances by the target model (CC-NM), the MIA accuracy changes as $0.551, 0.559, 0.502, 0.64, 0.58$, and 0.578. Further, when V contains all non-members incorrectly classified by the target model (IC-NM) the MIA accuracy differ as $0.79, 0.82, 0.76, 0.84, 0.70, 0.80$. From these results, it is conspicuous that by cherry-picking the data in V the overall accuracy of the MIA can be overstated or understated. Compared to selecting non-members randomly by using CC-NM based approach the MIA accuracy is understated by $10.91\%, 6.91\%$, and 4.18% for the over-fitted Location-30, Texas-100, and Purchase-100 models respectively. Further, by using IC-NM approach the membership inference accuracy is overstated by $25.9\%, 23.38\%$, and 20.2% for the MNIST, FMNIST, and Adult models which are well generalized. These results help us to understand two main aspects regarding MIAs. First, using aggregated membership inference results (i.e. accuracy) could be misleading when evaluating the attack models, mitigation strategies or simply when the model owners try to estimate the membership inference risk of a given model. Because by cherry-picking member/non-member data based on their classification correctness the membership risk can be artificially inflated or deflated. Therefore, better attack models need to be invented which are less sensitive to the classification correctness or the confidence score values. Next, we explore how the membership inference accuracy of different attack models change under CC-NM and IC-NM approaches for validation data (V) selection. Respectively for CONF, ENT, M-ENT, CuNN, LOSS, CC, TRANS, and ACT attack models the membership inference attack accuracy changes by 32.3%, 18.7%, 31.9%, 32.7%, 3.5%, 49.2%, 2%, and 4.6%. Out of the aforementioned methods LOSS, TRANS, and ACT membership attack models are the least sensitive to manipulation of membership validation data by CC-NM and IC-NM approaches. Among these, both the LOSS and the TRANS attacks have an overall poor membership inference performance. However, the ACT attack model significantly outperforms the rest of the attack models while being resistant to the above explained manipulation of the membership validation set.

Based on these results, we believe that claiming that the existing black-box MIAs can identify members used to train a target ML model is too strong. Rather, these attack models are capable of *determining whether a given data instance is similar or dissimilar to the underlying training dataset used to train a target ML model*. Nevertheless, for an extremely over-fitted ML model, this may lead to uncovering the true members. However, such a model would have minimum utility in practice due to poor generalizability. Hence, in the current state, these MIA models (that leverage confidence scores) pose more of a theoretical risk rather than a practical risk. Therefore, attack models like ACT are more preferred as it is capable of exploiting well generalized models. Moreover, identifying whether a given data instance is *similar* to the underlying training data distribution still entails privacy risks. For example, assume an adversary identified a tumor patient's record as a member of a target ML model trained to classify the dosage of a cancer drug given to terminally ill patients. This would indicate that the particular patient is likely to have a terminal cancer (instead of a benign tumor).

Table 4. Manipulating membership inference accuracy by selecting non-members for the balanced membership validation set based on a) correctly classified non-members (CC-NM) and b) incorrectly classified non-members (IC-NM).

DataSet (NM Type)	CONF	ENT	M-ENT	CuNN	LOSS	CC	TRANS	ACT
MNIST (CC-NM)	0.504	0.505	0.505	0.535	0.5	0.497	0.495	0.867
MNIST (IC-NM)	0.922	0.812	0.922	0.873	0.58	0.997	0.518	0.717
FMNIST (CC-NM)	0.542	0.542	0.542	0.539	0.5	0.5	0.484	0.827
FMNIST (IC-NM)	0.998	0.789	0.998	0.952	0.5	1	0.602	0.724
Adult (CC-NM)	0.478	0.508	0.478	0.476	0.5	0.437	0.5	0.645
Adult (IC-NM)	0.905	0.656	0.905	0.905	0.5	0.937	0.538	0.767
Location-30 (CC-NM)	0.724	0.727	0.723	0.554	0.553	0.497	0.5	0.853
Location-30 (IC-NM)	0.895	0.855	0.883	0.949	0.656	0.997	0.531	0.989
Texas-100 (CC-NM)	0.573	0.603	0.576	0.519	0.648	0.441	0.572	0.716
Texas-100 (IC-NM)	0.743	0.685	0.74	0.579	0.613	0.897	0.491	0.882
Purchase-100 (CC-NM)	0.577	0.579	0.575	0.521	0.518	0.488	0.505	0.86
Purchase-100 (IC-NM)	0.874	0.794	0.868	0.85	0.585	0.988	0.498	0.971

7   Conclusion

In this work, we propose a new black-box MIA model based on transfer attack and activation based outlier detection. The proposed attack model can outperform the existing black-box membership inference attack models with respect to both generalized and over-fitted DNN models thus challenging the conventional wisdom that states generalized models are immune to membership inference. Moreover, we highlight some significant weaknesses with respect to the existing confidence score based MIAs and systematically investigate them while showcasing that the newly proposed MIA model is immune to those. In future work, we plan to further study the proposed attack model with respect to different shadow data characteristics and available mitigation strategies for MIAs.

References

1. Chen, J., Jordan, M.I., Wainwright, M.J.: Hopskipjumpattack: a query-efficient decision-based attack. In: 2020 IEEE Symposium on Security and Privacy, SP 2020, San Francisco, CA, USA, 18–21 May 2020, pp. 1277–1294. IEEE (2020)
2. Choo, C.A.C., Tramer, F., Carlini, N., Papernot, N.: Label-only membership inference attacks. arXiv preprint arXiv:2007.14321 (2020)
3. Fredrikson, M., Jha, S., Ristenpart, T.: Model inversion attacks that exploit confidence information and basic countermeasures. In: Proceedings of the 22nd ACM SIGSAC Conference on Computer and Communications Security, pp. 1322–1333 (2015)
4. Hendrycks, D., Gimpel, K.: A baseline for detecting misclassified and out-of-distribution examples in neural networks. In: 5th International Conference on Learning Representations, ICLR 2017, Toulon, France, 24–26 April 2017. OpenReview.net (2017)
5. Humphries, T., et al.: Differentially private learning does not bound membership inference. arXiv preprint arXiv:2010.12112 (2020)
6. Jayaraman, B., Wang, L., Knipmeyer, K., Gu, Q., Evans, D.: Revisiting membership inference under realistic assumptions. arXiv preprint arXiv:2005.10881 (2020)
7. Jia, J., Salem, A., Backes, M., Zhang, Y., Gong, N.Z.: Memguard: defending against black-box membership inference attacks via adversarial examples. In: Proceedings of the 2019 ACM SIGSAC Conference on Computer and Communications Security, pp. 259–274 (2019)
8. Li, J., Li, N., Ribeiro, B.: Membership inference attacks and defenses in classification models. In: Proceedings of the Eleventh ACM Conference on Data and Application Security and Privacy, pp. 5–16 (2021)
9. Li, Z., Zhang, Y.: Membership leakage in label-only exposures. arXiv preprint arXiv:2007.15528 (2020)
10. Long, Y., et al.: Understanding membership inferences on well-generalized learning models. arXiv preprint arXiv:1802.04889 (2018)
11. Nasr, M., Shokri, R., Houmansadr, A.: Machine learning with membership privacy using adversarial regularization. In: Proceedings of the 2018 ACM SIGSAC Conference on Computer and Communications Security, pp. 634–646 (2018)

12. Nasr, M., Shokri, R., Houmansadr, A.: Comprehensive privacy analysis of deep learning: passive and active white-box inference attacks against centralized and federated learning. In: 2019 IEEE symposium on security and privacy (SP), pp. 739–753. IEEE (2019)
13. Rahman, M.A., Rahman, T., Laganière, R., Mohammed, N., Wang, Y.: Membership inference attack against differentially private deep learning model. Trans. Data Priv. **11**(1), 61–79 (2018)
14. Salem, A., Zhang, Y., Humbert, M., Fritz, M., Backes, M.: Ml-leaks: model and data independent membership inference attacks and defenses on machine learning models. In: Network and Distributed Systems Security Symposium 2019. Internet Society (2019)
15. Shokri, R., Stronati, M., Song, C., Shmatikov, V.: Membership inference attacks against machine learning models. In: 2017 IEEE Symposium on Security and Privacy (SP), pp. 3–18. IEEE (2017)
16. Song, L., Mittal, P.: Systematic evaluation of privacy risks of machine learning models. In: 30th {USENIX} Security Symposium ({USENIX} Security 2021) (2021)
17. Song, L., Shokri, R., Mittal, P.: Membership inference attacks against adversarially robust deep learning models. In: 2019 IEEE Security and Privacy Workshops (SPW), pp. 50–56. IEEE (2019)
18. Yeom, S., Giacomelli, I., Fredrikson, M., Jha, S.: Privacy risk in machine learning: Analyzing the connection to overfitting. In: 2018 IEEE 31st Computer Security Foundations Symposium (CSF), pp. 268–282. IEEE (2018)
19. Zhang, C., Bengio, S., Hardt, M., Recht, B., Vinyals, O.: Understanding deep learning requires rethinking generalization. In: 5th International Conference on Learning Representations, ICLR 2017, Toulon, France, April 24–26, 2017. OpenReview.net (2017)
20. Zhao, Y., Nasrullah, Z., Li, Z.: Pyod: A python toolbox for scalable outlier detection. J. Mach. Learn. Res. **20**(96), 1–7 (2019). http://jmlr.org/papers/v20/19-011.html

Webshell Detection Based on Explicit Duration Recurrent Network

Bailin Xie[1,2(✉)] and Qi Li[1,2]

[1] School of Information Science and Technology, Guangdong University of Foreign Studies,
Guangzhou, China
bailinxie@gdufs.edu.cn

[2] School of Cyber Security, Guangdong University of Foreign Studies, Guangzhou, China

Abstract. Webshell is a malicious script, it can be written in multiple languages. Through the webshell, attackers can escalate and maintain persistent access on compromised web applications. With the growth of the demand for interactive web applications, the webshell timely detection in web applications are essential to ensure security of web server. Although there are some existing methods for webshell detection, these methods need a large number of samples to achieve higher accuracy rate. In this paper, we proposed an new webshell detection method based on Explicit Duration Recurrent Network (EDRN). In this method, the opcode sequence of samples is considered as input using word2vec. Comparing with other Recurrent Neural Networks, such as LSTM and GRU, the experimental results illustrate that our model can achieve the better performance, especially when the training set is small.

Keywords: Webshell detection · Opcode sequence · Word2vec · Explicit Duration Recurrent Network

1 Introduction

With the rapid development of the Internet, interactive web applications are gradually becoming a window for people to exchange information. However, these applications are influenced by various web attacks. Although the applications under the protection of web vulnerability scanning become difficult to invade, due to the old habits of programmer in the development process are difficult to change, and there will inevitably be vulnerabilities in the system [1]. Attackers can make use of these common vulnerabilities such as SQL Injection (SQLi), Remote File Inclusion (RFI), or even use Cross-site Scripting (XSS) to upload a malicious file to the target web server. Through this malicious file which called webshell, attackers can obtain the execution authority of the server, such as executing system commands to acquire system information, etc., and finally achieve the purpose of controlling the web server.

According to the function and size, webshell can be roughly divided into three categories, namely Full Trojan, Mini Trojan and one-sentence Trojan. At present, researchers have proposed a lot of methods for webshell detection. Traditional webshell detection

W. Meng and M. Conti (Eds.): CSS 2021, LNCS 13172, pp. 55–65, 2022.
https://doi.org/10.1007/978-3-030-94029-4_4

methods are mostly based on rules to judge whether the file contains malicious keywords or malicious functions. However, these traditional methods usually have the problems of low accuracy, high false positive rate and high false negative rate. With the excellent performance of deep learning methods in text classification, researchers have also tried to apply deep learning technology to webshell detection. Existing studies have shown that the methods based deep learning can achieve better performance, comparing with the traditional detection methods.

The rest of the paper is organized as follows. Section 2 is the review of recently related research of webshell detection. In Sect. 3, we describe the detail of our detection method which based on EDRN. In Sect. 4, we conduct experiments for this method, the experiment results will be discussed with details. Finally, conclusion is given in Sect. 5.

2 Related Work

Webshell detection can be defined as a classification task which identifies the code as normal or malicious [15]. In order to implement the particular functions of webshell, its malicious code must have specific keywords and malicious functions. The existing webshell detection methods can be roughly divided into two types: static feature based detection and dynamic feature based detection [2].

The static feature based detection methods are mainly based on statistical analysis and feature matching. It means that checking from the file itself without executing the code. NeoPI [3] is a script written by Python, which detects webshell by calculating the statistical features of the file. Xu et al. [4] construct a webshell feature library and develop an ASP webshell search software based on this library. Tu et al. [5] perform statistical analysis on the malicious features that contained in files, and identify webshell based on optimal threshold values. With the rapid development of machine learning, some studies have begun to use machine learning for webshell detection. Hu et al. [6] proposed a webshell detection method based on Decision Trees, by extracting webpage features from three aspects of document attributes, basic attributes and advanced attributes. Zhu et al. [7] extract three features from PHP webshell, namely lexical features, syntactic features and abstract features. After that, they utilize Fisher score to select these features. Finally, support vector machine (SVM) is used to detect webshell. However, since some dangerous functions also appear in the normal code, these methods have a high false positive rate.

The dynamic feature based detection methods mainly focuses on the features when the code executing. Opcode is a part of computer instructions, which can directly reflect the running characteristics of a script regardless of whether the code is obfuscated or not. Therefore, some researchers have performed dynamic features detection by extracting the features of opcode sequences. Cui et al. [8] construct a two-layer model called RF-GBDT for PHP webshell detection, in which the first layer is used for dealing with the features of the opcode sequences based on the Random Forest classifier, and the detection is completed by combining the output of the previous layer and the statistical features of the PHP file through the Gradient Boosting Decision Tree classifier in the second layer. Ai et al. [9] propose a webshell detection model called WS-LSMR by integrating multiple classifiers of Logistic Regression, Support Vector Machine, Multi-layer

Perceptron and Random Forest. Furthermore, this model applied the Gini coefficient to perform feature selection and utilized a weighted voting method for classification. However, the influence of word order information on webshell detection is ignored in above methods. To alleviate this problem, some text classification models in neural network that can capture contextual information have been transferred to webshell detection. Fang et al. [10] propose an FRF-WD model for PHP webshell detection by combining fastText model and random forest algorithm, which uses the text features of opcode sequences and statistical features as input. Tian et al. [11] propose to utilize word2vec tool to denote the HTTP request as a matrix and apply convolutional neural network as a classifier. Since the turning points between sentences in text classification may lead to semantic reversal, Li et al. [12] propose to utilize the attention mechanism to pay attention to the inter-word information in webshell detection. In addition, Pan et al. [13] construct a feature matrix by combining static statistical features, text features of opcode sequences, and data execution characteristics for webshell detection. Zhao et al. [14] propose a webshell detection framework based on opcode sequences and taint analysis. However, these methods generally require a large number of samples to maintain high accuracy rate.

Different neural networks apply in different scenarios will make the model show different performance. Therefore, choosing a suitable neural network is crucial to the classification results. Due to the fact that there not many samples of webshell in the real environment and samples are difficult to obtain, we notice that the novel recurrent neural network called Explicit Duration Recurrent Networks (EDRN) proposed by Yu [16] is possible to drive the model converge faster and the number of training samples required by model can be reduced. In order to explore the effectiveness of EDRN in webshell detection, we propose a new webshell detection method based on the combination of word2vec representation and EDRN, which using PHP opcode sequences features of the webshells as input. To the best of our knowledge, this is the first time for EDRN to apply to webshell detection. The experimental results illustrate that compared with other Recurrent Neural Networks, EDRN can achieve better performance in webshell detection, especially when the training set is small, comparing with other Recurrent Neural Networks.

3 Model Architecture

In this section, we introduce a new webshell detection method based on EDRN which proposed by us in detail. The overall architecture of this detection model is shown in Fig. 1. This detection model consists of four parts: namely feature extraction, feature vectorization, EDRN, and dense layers. First, we need to obtain the opcode sequences of the PHP files by using a PHP extension called Vulcan Logic Disassembler (VLD) [17]. Second, each opcode sequence will be converted into a 100-dimensional vector by using word2vec. Finally, we train the webshell detection model by EDRN and make use of the dense layer to classify these PHP files.

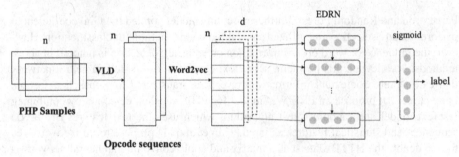

Fig. 1. The overall architecture of detection model.

3.1 Feature Extraction

When PHP gets a piece of code, the source code will be translated into the corresponding opcodes after lexical analysis, grammatical analysis and other stages, and then the Zend engine will execute these opcodes in sequence to complete the operation. VLD is an extension of PHP which hooks into the Zend engine and provides functionality to dump the internal representation of PHP scripts. An example is listed below, and the opcode sequences extracted by VLD is shown in Table 1.

```php
<?php
        @eval($_POST['cmd']);
?>
```

The opcode sequences of this example can be represented in the following form: ['BEGIN_SILENCE', 'FETCH_R', 'FETCH_DIM_R', 'INCLUDE_OR_EVAL', 'END_SILENCE', 'RETURN'].

Therefore, the opcode sequence of each file can be represented as O_n:

$$O_n = [W_{n,1}, W_{n,2}, \ldots, W_{n,m}] \tag{1}$$

where $W_{n,j}$ denotes the jth opcode in the nth opcode sequence, m stands for the number of opcodes in a opcode sequence.

3.2 Feature Vectorization

After obtaining the opcode sequences, we convert these opcode sequences into vectors representation. Compared with word2vec [18], the vectors obtained by one-hot encoder usually have a high dimension. At the same time, since one-hot encoder ignores the relationship between each opcode, it cannot represent the features of the code better. For these reasons, we adopt the CBOW algorithm of word2vec to achieve this operation. The opcode sequence O_n vectorized by word2vec can be represented as V_n:

$$V_n = [V_{n,1}, V_{n,2}, \ldots, V_{n,m}] \tag{2}$$

where $V_{n,j}$ denotes the vector form of the jth opcode in the nth opcode sequence.

Table 1. Dump of the opcode sequences.

#	OPCODE
1	BEGIN_SILENCE
2	FETCH_R
3	FETCH_DIM_R
4	INCLUDE_OR_EVAL
5	END_SILENCE
6	RETURN

3.3 EDRN

Similar to common Recurrent Neural Networks, the Explicit Duration Recurrent Network also includes input gate, output gate and forget gate. The distinguishing characteristic of this new recurrent neural network is that all the states and substates are no longer the same. And it can distinguish which substate is corresponding to hidden node. We suppose the EDRN has M states and each state consists of D substates. Let the observation sequence at time t is denoted as x_t, the calculation formulas for the forget gate G_{fg}, input gate G_{in}, tanh gate G_{th} are as follows:

$$G_{fg} = \sigma\left(h_{t-1}A_{fg} + x_tB_{fg} + b_{fg}\right) \tag{3}$$

$$G_{in} = \sigma(h_{t-1}A_{in} + x_tB_{in} + b_{in}) \tag{4}$$

$$G_{th} = tanh\left(C_{t-1}A_{pt} + C_{t-1}(:, D)A_{th} + x_tB_{th} + b_{th}\right) \tag{5}$$

Then the memory cell C_t for the time step t can be updated by the following formulas:

$$C_t = C_{t-1} * G_{fg} + G_{th} * G_{in} \tag{6}$$

Thus, the output gate G_{ot} and the hidden state h_t for the time step t can be calculated by the following formulas:

$$G_{ot} = \sigma(C_t(:, D)A_{ot} + x_tB_{ot} + b_{ot}) \tag{7}$$

$$C_t^{**} = tanh(C_t) * G_{ot} \tag{8}$$

$$h_t(:) = \sum_{d=1}^{D-1} C_t^{**}(:, d) + C_t^{**}(:, D)A_{st} \tag{9}$$

where σ represents the point-wise non-linear activation functions of logistic sigmoid, A, B. and b are trainable parameters, ":" denotes all the states. Details of the EDRN can be found in [16].

4 Experiment

This section, we conducted experiments tevaluate the performance of EDRN in webshell detection and compared it with other Recurrent Neural Netwks, such as LSTM. In our experiments, we only focus on the PHP type samples.

4.1 Dataset

We collected 2229 PHP webshell samples [19–22] which come from github, and 3348 white samples which come from several open source CMS projects [23–25]. After the data processing stage, the total number of samples, which the opcode sequence features are successfully extracted and retained for the experiment, is 5353. The number of webshell samples is 2009, The number of white samples is 3344. In order to prevent the imbalance of sample distribution in training dataset and testing dataset from affecting the experimental results, we always keep a similar distribution proportion of white samples and webshell samples in the process of the experiment.

4.2 Evaluation Criteria

Unlike other classification tasks, when evaluating the webshell detection model, we need pay more attention to whether the webshell samples are correctly identified or not. Therefore, we adopt the accuracy rate to evaluate the ability to identify webshell and white samples, the recall rate to reflect the false negative rate (FNR) of the model, and the F1 value to evaluate the robustness of the model. The confusion matrix is shown in Table 2.

Table 2. Confusion matrix.

	Webshell	White
Webshell	True Positive (TP)	False Negative (FN)
White	False Positive (FP)	True Negative (TN)

Then the calculation formulas for accuracy rate, recall rate and F1 value are as follows:

$$Accuracy = \frac{TP + TN}{TP + FN + FP + TN} \tag{10}$$

$$Recall = \frac{TP}{TP + FN} \tag{11}$$

$$F1 = \frac{2 * Precision * Recall}{Precision + Recall} \tag{11}$$

4.3 Experiment Result

Cording to the overall situation of all samples, we convert each sample into a 100-dimensional vector. Then, we respectively extract different proportions of samples from the data set as the training set, and the rest as the testing set. As shown in Fig. 2, we found that when the epoch is set as 15, EDRN, LSTM and GRU can achieve the best and comparable results. Therefore, we set the units = 128, epoch = 15 and batch_size = 32 as the default parameters for all experiments.

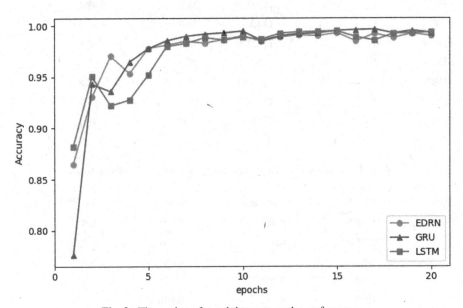

Fig. 2. The setting of epoch impacts on the performance.

In addition, in order to make the total number of EDRN parameters not greater than 117,377 of LSTM, we explore the impact of different parameter settings on EDRN by setting different units and duration for EDRN. The experimental results are provided in Table 3. These results demonstrate that when we set units = 35 and D = 6 for EDRN, it performed best. Thus, we set units = 35 and D = 6 for EDRN in our experiments by default, where D is the value of duration.

In our experiments, we explore the performance of EDRN for webshell detection under different size of training datasets, and compare the results with other three recurrent neural networks, including RNN, GRU and LSTM.

Table 4 shows the experimental results of EDRN, LSTM, GRU and RNN under different size of training datasets, where "size" represents the proportion of training dataset samples to the whole dataset.

Table 3. The impact of different parameter settings on EDRN.

Model	D	Units	Params	Accuracy (%)	Recall (%)
LSTM		128	117,377	99.1	98.5
GRU		128	88,065	99.1	98.6
EDRN	2	77	115,065	98.9	98.6
EDRN	3	59	116,998	98.6	98.8
EDRN	4	48	117,073	98.7	98.1
EDRN	5	40	114,841	98.0	95.7
EDRN	6	35	116,026	99.2	98.6
EDRN	7	31	116,220	98.9	97.6
EDRN	8	28	117,181	98.0	98.8
EDRN	9	25	114,951	98.3	98.3
EDRN	10	23	115,668	99.0	98.1

Table 4. Experimental results under the different size of training datasets.

Size	Model	Accuracy	Recall	F1
0.2	EDRN	**0.981**	**0.971**	**0.974**
	LSTM	0.958	0.951	0.954
	GRU	0.968	0.959	0.958
	RNN	0.866	0.760	0.811
0.4	EDRN	**0.982**	**0.970**	**0.974**
	LSTM	0.969	0.964	0.959
	GRU	0.974	0.961	0.966
	RNN	0.892	0.800	0.850
0.6	EDRN	0.988	**0.989**	0.985
	LSTM	**0.989**	0.984	**0.986**
	GRU	0.987	0.982	0.983
	RNN	0.902	0.801	0.861
0.8	EDRN	**0.992**	0.986	**0.989**
	LSTM	0.991	0.985	**0.989**
	GRU	0.991	**0.986**	0.988
	RNN	0.908	0.942	0.888

It can be seen from the Table 4 that the results of RNN are the worst. When only 20% of the samples are used as the training dataset, the accuracy rate of EDRN can reach an astonishing 0.981, and its recall rate is 0.971, which are obviously higher than that of GRU and LSTM. As the proportion of training dataset continues to increase, the accuracy rates of EDRN, GRU and LSTM are gradually increasing. When it reaches 80%, their results are comparable. Overall, the accuracy rate of EDRN is always above 0.98, which illustrates that EDRN has a good capability to classification of webshell.

Furthermore, by comparing the recall rates and F1 values of different models under different sizes of training datasets, we find that only when 80% of the samples are used as the training dataset, LSTM and GRU can achieve the comparable results with EDRN. But in other cases, EDRN always shows the best performance. It demonstrates that EDRN can more effectively distinguish the webshell and normal code.

Figure 3 is a comparison chart of the results of EDRN, LSTM and GRU. It can be intuitively seen from Fig. 3 that EDRN can achieve higher accuracy rate, recall rate and F1 value with a small training dataset in webshell detection. As the training set increases, the results of LSTM and GRU are gradually comparable to EDRN. It demonstrates that EDRN is more suitable for webshell detection.

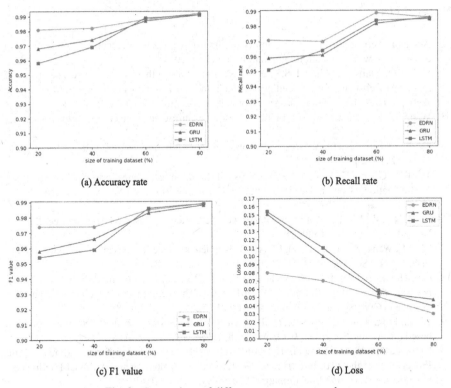

(a) Accuracy rate

(b) Recall rate

(c) F1 value

(d) Loss

Fig. 3. Comparison of different recurrent networks.

5   Conclusion

In this paper, we proposed a new webshell detection model which based on Explicit Duration Recurrent Network and word2vec. Then, we explore the effectiveness of EDRN in webshell detection and compare the results with other recurrent networks. The experimental results demonstrate that our method is better than the other recurrent networks, especially when the training dataset is small. This is what is pursued in the actual environment.

At present, we only used the opcode sequence features of the PHP samples in our experiment. In the future, we will incorporate more features on the basis of this method.

Acknowledgments. This work is supported by the Guangdong Basic and Applied Basic Research Foundation (Grant No. 2018A0303130045), the Science and Technology Program of Guangzhou (Grant No. 201904010334).

References

1. Acunetix Web Application Vulnerability Report 2020. https://www.acunetix.com/white-pap ers/acunetix-web-application-vulnerability-report-2020/
2. Starov, O., Dahse, J., Ahmad, S.S., et al.: No honor among thieves: a large-scale analysis of malicious webshells. In: Proceedings of the 25th International Conference on World Wide Web, pp. 1021–1032. Association for Computing Machinery, New York (2016)
3. Web Shell Detection Using NeoPI. https://resources.infosecinstitute.com/topic/web-shell-det ection/
4. Xu, M., Chen, X., Hu, Y.: Design of software to search ASP webshell. Procedia Eng. **29**, 123–127 (2012)
5. Tu, T.D., Guang, C., Guo, X., et al.: Webshell detection techniques in web applications. In: 5th International Conference on Computing, Communications and Networking Technologies (ICCCNT), pp. 1–7. IEEE, New York (2014)
6. Hu, J., Xu, Z., Ma, D., et al.: Research of webshell detection based on decision tree. J. Netw. New Media **1**(6), 15–19 (2012)
7. Zhu, T., Weng, Z., Fu, L., Ruan, L.: A webshell detection method based on multiview feature fusion. Appl. Sci. **10**(18), 6274 (2020)
8. Cui, H., Huang, D., Fang, Y., et al.: Webshell detection based on random forest–gradient boosting decision tree algorithm. In: 2018 IEEE Third International Conference on Data Science in Cyberspace (DSC), pp. 153–160. IEEE, New York (2018)
9. Ai, Z., Luktarhan, N., Zhao, Y., et al.: WS-LSMR: malicious web shell detection algorithm based on ensemble learning. IEEE Access **8**, 75785–75797 (2020)
10. Fang, Y., Qiu, Y., Liu, L., et al.: Detecting webshell based on random forest with fasttext. In: Proceedings of the 2018 International Conference on Computing and Artificial Intelligence, pp. 52–56. Association for Computing Machinery, New York (2018)
11. Tian, Y., Wang, J., Zhou, Z., et al.: CNN-webshell: malicious web shell detection with convolutional neural network. In: Proceedings of the 2017 VI International Conference on Network, Communication and Computing, pp. 75–79. Association for Computing Machinery, New York (2017)
12. Li, T., Ren, C., Fu, Y., et al.: Webshell detection based on the word attention mechanism. IEEE Access **7**, 185140–185147 (2019)

13. Pan, Z., Chen, Y., Chen, Y., et al.: Webshell detection based on executable data characteristics of PHP code. Wirel. Commun. Mob. Comput. **2021**, 5533963 (2021)
14. Zhao, J., Lu, Y., Wang, X., et al.: WTA: a static taint analysis framework for PHP webshell. Appl. Sci. **11**(16), 7763 (2021)
15. Yang, W., Sun, B., Cui, B.: A web shell detection technology based on HTTP traffic analysis. In: Barolli, L., Xhafa, F., Javaid, N., Enokido, T. (eds.) Innovative Mobile and Internet Services in Ubiquitous Computing. IMIS 2018. Advances in Intelligent Systems and Computing, vol. 773, 336–342. Springer, Cham (2019). https://doi.org/10.1007/978-3-319-93554-6_31
16. Yu, S.: Explicit duration recurrent networks. IEEE Trans. Neural Netw. Learn. Syst. 1–11 (2021)
17. Vulcan Logic Dumper. http://pecl.php.net/package/vld
18. Mikolov, T., Chen, K., Corrado, G., Dean, J.: Efficient estimation of word representations in vector space. arXiv preprint arXiv:1301.3781 (2013)
19. Ysrc-webshell-sample. https://github.com/ysrc/webshell-sample
20. X17dev-webshell. https://github.com/xl7dev/WebShell
21. JohnTroony-php-webshells. https://github.com/JohnTroony/php-webshells
22. BlackArch-webshells. https://github.com/BlackArch/webshells
23. Cratfcms. https://github.com/craftcms/cms
24. WordPress. https://github.com/WordPress/WordPress
25. Yii2. https://github.com/yiisoft/yii2

A Practical Botnet Traffic Detection System Using GNN

Bonan Zhang$^{(\boxtimes)}$, Jingjin Li, Chao Chen, Kyungmi Lee, and Ickjai Lee

College of Science and Engineering, James Cook University,
Townsville, QLD 4811, Australia
bonan.zhang@my.jcu.edu.au

Abstract. Botnet attacks have now become a major source of cyber-attacks. How to detect botnet traffic quickly and efficiently is a current problem for most enterprises. To solve this, we have built a plug-and-play botnet detection system using graph neural network algorithms. The system detects botnets by identifying the network topology and is very good at detecting botnets with different structures. Moreover, the system helps researchers to visualise which nodes in the network are at risk of botnets through a graphical interface.

Keywords: Botnet detection · Graph neural network · Machine learning

1 Introduction

A botnet is a network of computers that malicious attackers have controlled. Botnets are often used to send a large number of spam packets to an attack target at the same time to make the target less productive or even to stop serving normal users. The harm of botnet attacks is pervasive worldwide, and it has also caused substantial security risks and economic losses. According to the results of a global survey conducted by the Ponemon Institute, the number one category of botnet attacks is the denial of service (DoS) and distributed denial of service (DDoS) attacks. These two types of attacks account for more than 55% of all cybercrimes each year and are the most influential cyber crimes. Furthermore, world-renowned company Akamai predicts that if a data centre is the target of a DDoS attack, there is a 25% chance that it will be attacked again within three months. There is a 36% chance that will be the target of attack again within a year [10].

In order to avoid detection, attackers have been developing and improving the architecture of botnets in recent years. There are currently two main architectures, the Client-server model and Peer-to-peer model [8]. Traditional botnets use a client-server model in which the botnet nodes operate mainly through Internet relay chat networks, domain names or websites. In this structure, a botnet herder controls the individual bot nodes by sending commands to the server to carry out attacks, and the botnet nodes send the results of the attacks back to the botnet herder.

© Springer Nature Switzerland AG 2022
W. Meng and M. Conti (Eds.): CSS 2021, LNCS 13172, pp. 66–78, 2022.
https://doi.org/10.1007/978-3-030-94029-4_5

In addition to traditional botnet architecture, some newer ones use peer-to-peer (P2P) botnet architecture. The behaviour of those two types of botnets is the same. The difference is that P2P botnets do not need to rely on a central server to communicate. They can share resources through interconnected nodes. Communication via P2P networks can be better prevented from being detected. These bots can use electronic signatures and other methods. Therefore only those that can obtain the private key can control the botnet. Using P2P networks to achieve botnet attacks can also be better than centralized botnets, which can be any single point of failure.

So far, there are three methods to detect and identify botnet traffic. The first is to detect botnet nodes through traffic patterns [6]. This detection technique based on the content of network traffic is often used for intrusion detection of systems. It first analyzes the content characteristics of the network traffic. It then matches the attributes of the botnets found in the network to determine whether it is in the network with such a botnet. The detection technology needs to maintain a feature library for fast matching and detection accuracy. This method can be easily deployed, which makes it widely used by security personnel to detect botnets. The second approach requires a priori knowledge to stop botnet attacks. This method blocks access to the botnet through DNS blacklisting, thus avoiding being kidnapped by the botnet [1]. However, this method requires the detection system to be deployed on the backbone ISP node, which threatens the private information on the Internet. The third method is to detect botnet traffic by identifying the specific topology of the botnet. Because botnets often have unique topology structures, such as mixing rate, the number and size of connection graph components, and other characteristics. Starting from the topology structure, some studies have found that centralized botnets have a unique hierarchical structure, while P2P botnets have a faster mixing rate. The faster mixing rate is due to the need for P2P structured botnets to share and transmit information quickly. A more detailed introduction of the three detection methods will be discussed in the "Related Work" section.

In recent years, in response to various network security issues, various machine learning models have been widely used and have achieved great success. For example, it is used for malware detection [14], network threat/incident detection [12,16] and software vulnerability detection etc. [9,11,15]. The machine learning model has satisfactory results in terms of accuracy and speed of detection. In addition, some studies have shown that using machine learning models to detect botnet traffic is also very effective. However, current botnet detection models often require a large amount of network traffic information, including traffic port numbers, traffic duration, traffic protocols and other information for training. The method of judging botnet traffic based on traffic information analysis has many limitations in real applications. Attackers can also avoid detection by various methods, such as encrypting traffic and modifying traffic information. Other methods that use topology features [13] to identify botnet traffic require a lot of human labour. In real-world applications, constant modification of parameters and multiple pre-filtering steps make the detection efficiency too low. To solve the various shortcomings of the existing methods, in this paper, we use a multi-layer GNN (Graph Neural Network) model to realize the automatic recognition of the network topology. The GNN model can efficiently and automatically identify the relationship between nodes in the network. Because in each layer of the GNN model, each node passes messages to its neighbours to exchange information and update its state. Then after the multi-layer GNN model, the data of each node will contain the information of its neighbouring nodes, and this associated information can help the algorithm to identify the botnet network topology more efficiently.

In this paper, we present a plug-and-play system for detecting botnet traffic, which was not included in previous studies. Our approach addresses the challenge of how SMEs can deploy botnet defences. Our system makes it possible to detect botnet traffic in the LAN. We have implemented an anomalous traffic detection platform with extended capabilities that can effectively detect botnet traffic from various architectures. Only the source and target IPs of the traffic are required in the detection system to be able to detect botnet. The system does not require detailed traffic information such as port information, traffic duration, etc., ensuring the privacy of the data. We experimented and tested our system's ability to detect botnets of both architectures. The experiments showed that our detection rate exceeded 97% for both architectures of botnets. In our experiments, our system was able to efficiently classify nodes and give classification results quickly even for large datasets. And the system's visualisation module allows users to visualise the topology of incoming traffic and discover which nodes on the LAN are communicating with the botnet. The system is easy to deploy and offers one-click execution.

The structure of this paper is as follows: in Sect. 2, we present the latest techniques currently used to detect botnets. In Sect. 3, we describe in detail the structure of our system and how the various modules are implemented. In Sect. 4, two botnet structures are selected as case studies to test and analyse the performance of our system in detail. Section 5 is our summary of this paper and the direction we envisage for the future of the optimization system.

2 Related Work

Botnet detection can be divided into three main streams based on the algorithms they are designed for. Each of these three streams has its advantages and disadvantages. Therefore, we will discuss and provide an insight into which streams to use as a guidance for specific scenarios.

2.1 From Traffic Pattern

Detection of botnets by traffic is a very effective method. BotSniffer is one of the methods to detect botnets by their traffic [5]. BotSniffer is based on the idea that the instructions and messages delivered by a botnet to each node should be similar. The method consists of two parts: the monitoring engine and the correlation engine. The monitoring engine will detect network traffic and pass suspicious traffic logs to the next component for analysis. The correlation engine will cluster the traffic to find similar sets of traffic and thus find botnet clusters. Another high precision method, developed by Barlos et al. in the year 2016 [2], compiled a set of statistical features and set up as a reference for the analyze component to detect the malware through inspecting the network traffic for similar information.

Identifying botnets through network traffic packet information can achieve high accuracy and is easy to deploy. However, this detection can easily be evaded by an attacker. A research paper published by Nedim et al. in 2014 showed that attackers could avoid detection by using approximate algorithms and automatic inference algorithms without obtaining any insider information about the detection system. This can make the accuracy of the detector drop from 100% to 28%–33% [7]. The paper published by Gu et al. in 2008 also showed that botnets could also evade traffic detection through encryption channels and other methods [13].

2.2 Use DNS Black List

Another detection method requires a priori knowledge of the botnet. A blacklist of network domains is obtained to analyse whether system access is related to a botnet. And by disabling communication between the botnet nodes and the host, botnet attacks can be avoided. This type of defence is a passive defence. However, the network domain blacklist used in this detection method is generally public and cannot be updated in time, so it is easy for attackers to evade detection.

2.3 From Traffic Topology

Identifying botnets by network topology has become a new trend, as recent research has found that botnets of both architectures have specific topology. Especially in recent years, with the development of graphical neural network technology, botnet identification by network topology is becoming more and more accurate. For the centralized botnet structure, their topology structure has an apparent hierarchical structure. For P2P botnets, in order to perform malicious attacks, they need to share and disseminate information more efficiently, so they have a higher mixing rate than normal network structures. Nagaraja et al. proposed a method to detect botnets based on the topology characteristics of P2P botnets with a high mixing rate [13]. Collins & Reiter also proposed in their paper a method that can detect botnets based on the number and size of connected graph components [3]. But it is not difficult to see that due to the enormous network traffic, it is still challenging to distinguish the topology of each traffic. This method requires a lot of workforce to mark topology features and multiple filtering steps to narrow the scope of suspicion. And in the practice of this method, it is necessary to continuously adjust the model parameters and so on according to different data. In the study of Jia Wei et al. in 2020, a method of using Graph Neural Network (GNN) to identify network topology from massive network traffic was proposed [17].

3 GNN Based Botnet Traffic Detection System

3.1 System Overview

To quickly and accurately identify bot traffic in network traffic, we propose the abnormal traffic detection system to encode, extract features and visualize traffic data, and implement the task of classifying nodes through GNN models. The system can be divided into three parts, the data processing module, the model classification module, and the result visualization module. The structure of this system is shown in the diagram below.

Fig. 1. The structure of abnormal traffic detection system

3.2 Data Processing Module

In this module, we will transform traffic data into graph data. Converting traffic data into graph representations is a standard method used to monitor, analyze and visualize network traffic. Once converted into graph data, they can be accepted by the underlying machine learning (ML) algorithms. More importantly, the transformation into graph data will allow data to express many implicit features that will help us in our classification work. It can help the system better analyze and identify the relationships between nodes and thus better classify them.

The system can accept input data in two formats. One contains the source and destination IP addresses of the traffic. The other format of input data will contain, in addition to the IP address, label information on whether the traffic message is botnet traffic. When transformed into graph data, each node in the graph will correspond to an entity with a different IP address, and the system will follow the following two rules to filter the data.

Node filtering: If an IP node has only network traffic that communicates with itself, the node will be filtered. This filtering is done for two reasons: Firstly, if a node has no traffic with other nodes, it means that the probability of the node becoming a botnet node is very low. Secondly, such filtering will reduce the number of nodes we need to classify and increase the efficiency of our system.

Edge Filtering: First of all, the edges of the graph constructed in our system have no direction. For a data input format with only source and destination IPs, if there is communication traffic between node a and node b, the system adds an edge between node a and node b. If there is any subsequent communication traffic between nodes a and b, the system ignores this traffic. The conversion to graph data is the same for datasets containing label information, except for an additional step to record the node classification. If all communication between two nodes is normal, both nodes will be classified as normal. However, if any of the communications between the two nodes are labelled as botnet communications, both nodes will be classified as botnet nodes.

The data set will be transformed by the above module into a graph pattern that the GNN model will recognise and handed over to the next module to classify the nodes.

3.3 Detection Module with GNN

Our system is designed to detect botnets by identifying their specific topology. In this work, our system automatically identifies topology features of the network (i.e. communication patterns) and classifies nodes through a trained graph neural network (GNN) [17]. This GNN model consists of an input layer output layer and ten hidden layers. The model updates node features at each layer by using convolutional neural networks, where nodes update their state and exchange information by passing messages to their neighbouring nodes. After 12 layers, a node will contain information about its neighbouring nodes within 12 hops. The model structure of GNN is shown in the figure below.

From the Fig. 2, we can see that the model uses 32 vectors to represent the features of the nodes collectively. In the output layer of the model, the final classification is achieved through a softmax function. For a multilayer graph convolutional neural network its layer-to-layer equation is:

$$X^{(l)} = \sigma(\bar{A} X^{(l-1)} W^{(l)})$$

where $\bar{A}$ is calculated as

$$\bar{A} = D^{-\frac{1}{2}} A D^{-\frac{1}{2}}$$

In this equation A is the adjacency matrix of the graph, D is the degree matrix of the graph nodes, and W is the learnable matrix at layer l. σ is the non-linear activation function ReLU. The calculation of $\bar{A}$ in this equation depends on the degree of the source node and the degree of the target node. This model makes changes to the calculation of $\bar{A}$: $\bar{A} = D^{-1} A$. By adjusting the way $\bar{A}$ is calculated, the model can control how the features of neighbouring nodes are normalised before aggregation. In contrast to the standard calculation, $\bar{A}$ in the model is only calculated in relation to the degree of the source node. This allows a node to have an equal proportion of information passed to it by its neighbouring nodes, resulting in a random walk. The reason for the model to adjust the calculation of $\bar{A}$ is to better identify botnets with P2P structures. This is because the information about the target node does not need to be taken into

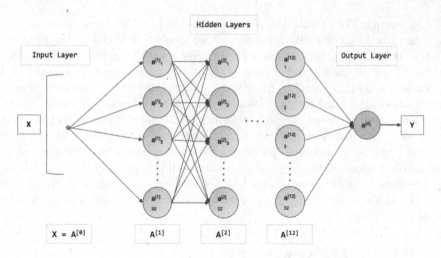

Fig. 2. The structure of GNN model

account when calculating the mixing rate. The probability of a node propagating to each of its neighbouring nodes should be the same.

3.4   Visualisation Module

This module is responsible for visualising the model classification results to the user. An online web service for anomaly detection using Django has been built to enable users to detect unusual traffic. The module's workflow is divided into the following three steps: user uploads files, display of traffic statistics and presentation of model classification results. In the file upload step, the user needs to upload a file that includes the source and destination IP of the traffic. It is optional to include the traffic label information in the file. After the user has finished uploading the file, the system will first count the traffic information in the file. The dataset's number of nodes, edges, abnormal nodes, and abnormal edges are presented to the user in a table format. Once the system has completed the classification process, if the user uploads a file that contains label information for the traffic, the system will compare the model's predicted results with the actual label results. Model classification accuracy, false positive rate, false negative rate, recall, precision, F1 values and auroch values will also be presented to the user in tabular form. To be able to visualise the predictions, we use Echarts to display the traffic topology. The black nodes are the normal nodes, and the red nodes are those predicted to be botnets. By clicking on the red nodes in the diagram, the researcher can easily find out which normal nodes are connected to the botnet nodes. The Fig. 3 shows the graphical interface of our system.

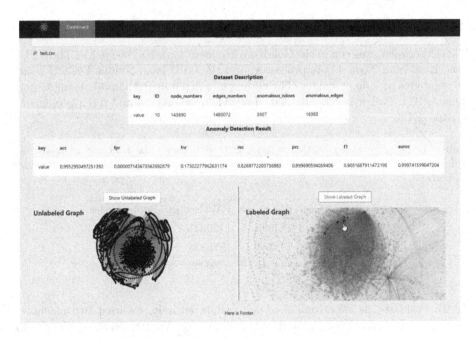

Fig. 3. The graphical interface of our system

4   System Evaluation and Case Studies

In this section, we will evaluate our model in terms of both the effectiveness and efficiency of the system. The dataset used to test the system was a synthetic botnet traffic dataset collected by Jiawei et al. [17]. This dataset uses all IP traces collected by CAIDA's monitor in 2018 as background traffic. In this dataset, after aggregating the traffic graphs, they are split into subnetwork traffic. The subnet traffic is also selected for training and testing in this experiment, due to the fact that the subnet traffic matches the topology of network traffic collected by an enterprise or research institution. The botnet is generated by randomly selecting a subset of the background traffic as botnet nodes and embedding the botnet topology into this subset of nodes. This dataset embeds the decentralised botnet P2P captured by Garcia et al. [4] in 2011 into the background traffic.

Both the P2P structured botnet and the C&C structured botnet dataset used in this project contain 960 graphs. Each of these graphs contains 3K botnet nodes. We randomly split the dataset into a training set, a validation set and a test set in the ratio of 8:1:1. The data used in this project has been desensitised, and all IP addresses are represented in the form of serial numbers. The distribution of the dataset and the statistics of nodes and edges are shown in Table 1.

4.1 Experiment Settings and Environment

This experiment was run in the Colab environment launched by google. The CPU used is an Intel Xeon 4-core processor, and the GPU is an Nvidia T4GPU with 16 gigabytes of video memory. The code was written in Python 3.7, using Cuda version 10.1, PyTorch version 1.4.0, torchvision version 0.5 and torch-geometric version 1.4.3.

Table 1. Botnet dataset statistics for a P2P structure. The background traffic is different for each graph, but the dataset embeds the same number of botnet nodes for each graph.

Data split	Graphs	AVG nodes	AVG edges	Botnet nodes
Train	768	143895	729231	3000
Val	96	143763	728934	3000
Test	96	144051	730089	3000

To evaluate the effectiveness of our system entirely, we used two machine learning algorithms, XGBoost and logistic regression. We trained the model using two dataset feature extraction methods. The first data extraction method is to extract the graph embedding features of the nodes directly using the GNN trained by the system. The second type of data extraction is the use of traditional graph features. The graph features include the node's degree, the maximum, minimum and average values of neighbouring nodes' degrees. On the other hand, we evaluate the system's efficiency by adjusting the size of the test set.

4.2 Evaluation Metric

When evaluating the model performance, we can observe that the data set is extremely unbalanced. In reality, the number of normal nodes should be much higher than the number of botnet nodes. In the dataset we used, the number of botnet nodes is the only 2.1% of the total number of nodes. In this case, evaluating the model's accuracy is not very meaningful for assessing the model performance. On the other hand, the system's primary purpose is to find all the botnet nodes in the network and improve the classification precision. So when evaluating the system's performance, our focus is on the recall and false positive rate of the system. To complement this, we also include the precision and F1-measure values of the system as metrics to evaluate performance.

4.3 Case Study 1: P2P Botnet

In this section, we focus on evaluating the performance of our system when classifying botnets with P2P structures. Table 2 shows the prediction results of our system and the two machine learning algorithms when trained with graph-embedded data.

Table 2. Results of botnet classification for P2P structures

Algorithm	Metric			
	Recall	FPR	Precision	F-measure
GNN model	98.4	0.00	99.4	98.91
XGBoost	98.5	37.6	5.4	10.2
Logistic regression	44.5	27.3	3.4	6.3

The results in Table 2 show that our system has a high botnet detection rate and guarantees a very low false positive rate. The XGBoost algorithm has a high botnet detection rate, but its false positive rate is higher than our system. Although its false positive rate is low for the logistic regression algorithm, it has a much lower botnet detection rate than the other two algorithms. And due to the low false positive rate of our system, the precision and F1-measure values of the system are also high. In contrast, the precision and F1-measure values of both the XGBoost and logistic regression algorithms are low.

Table 3. Performance of algorithms trained using traditional features

Algorithm	Metric			
	Recall	FPR	Precision	F-measure
GNN model	98.4	0.00	99.4	98.91
XGBoost	99.1	98.9	2.2	4.3
Logistic regression	91.4	91.3	2.2	4.3

Table 3 shows the performance of the two machine learning algorithms when trained using traditional graph features. As can be seen from the table, both the XGBoost algorithm and the logistic regression algorithm have high botnet detection rates. However, both algorithms have a high false positive rate. Similarly, both have low precision and F1 values. This shows that using only 2-hop information to identify botnets is not enough.

Figure 4 shows the variation in the system running time for different numbers of data sets. Where the horizontal coordinate is the number of graphs contained in the dataset used for testing. The vertical coordinate is the time taken to classify the nodes in the system. We can see that the running time of the system increases in an essentially linear fashion. In the case of classifying thirteen million nodes, the system can still complete the classification of nodes within the 40s. This classification is efficient enough to meet the needs of most companies and research institutes.

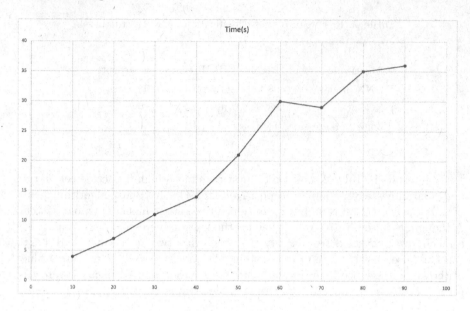

Fig. 4. System runtime with different number of graph (P2P)

4.4   Case Study 2: C&C Botnet

The classification performance of our system and the two machine learning algorithms in identifying botnets with C&C structures is shown in Table 4. It is clear that the GNN model still has a high recognition rate and a low false positive rate in identifying botnets under this structure. For the XGBoost algorithm, its performance does not change much from that when identifying P2P structures. While the LR algorithm has improved detection rates, it has also increased its false positive rate. It shows that our system performs well in identifying botnets of both structures and is able to identify the botnet traffic in the traffic effectively.

Table 4. Results of botnet classification for C&C structures

Algorithm	Metric			
	Recall	FPR	Precision	F-measure
GNN model	96.4	0.00	99.8	98.0
XGBoost	96	31.5	6.3	11.8
Logistic regression	62.2	43.2	3.1	5.9

In the same way as Fig. 4, Fig. 5 shows the time taken by our system to process a botnet with a C&C structure. We can see that the system runtime tends to increase linearly with the size of the data set. However, the system takes slightly more processing time to handle networks with a C&C structure than a P2P network structure. This may be due to the fact that many of the nodes in the P2P structured network are only connected to themselves for communication, resulting in many nodes being filtered.

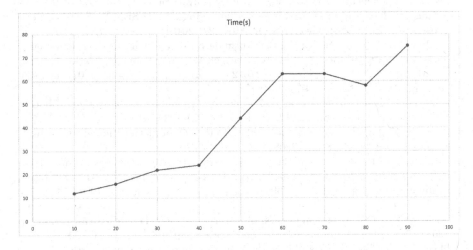

Fig. 5. System runtime with different number of graph (C&C)

5   Conclusion and Future Work

In this work, we propose a plug-and-play system for detecting anomalous traffic in the network. Experiments show that our system is effective and has a high botnet detection rate compared to other machine learning algorithms. Our system is also able to identify botnet traffic for both structures effectively. The system has a high classification efficiency and is able to meet the needs of most companies and research institutes.

For future research directions in this project, we hope to expand the functionality of the ontology. Firstly, in this project, we only identified the botnet traffic in the network topology diagram, but we did not test whether other anomalous traffic could be identified. Such as worms or viruses attack traffic. To further improve the accuracy of the predictions, in future work, we can use additional traffic information such as packet size and port numbers as well as domain names or DNS blacklists in addition to topology information. On the other hand, In the current case, in order to speed up the training, we use a topology map that is undirected and has no weights on the edges. Moreover, the network topology we use for training is a static graph, from which we can only see that there is communication between two IP nodes. Still, we cannot obtain information such as the frequency of communication between the two IP nodes. We can also modify this in future work to improve our model predictions.

References

1. Alieyan, K., Almomani, A., Anbar, M., Alauthman, M., Abdullah, R., Gupta, B.B.: DNS rule-based schema to botnet detection. Enterp. Inf. Syst. **15**(4), 545–564 (2021)
2. Bartos, K., Sofka, M., Franc, V.: Optimized invariant representation of network traffic for detecting unseen malware variants. In: 25th {USENIX} Security Symposium ({USENIX} Security 16), pp. 807–822 (2016)
3. Collins, M.P., Reiter, M.K.: Hit-List worm detection and bot identification in large networks using protocol graphs. In: Kruegel, C., Lippmann, R., Clark, A. (eds.) RAID 2007. LNCS, vol. 4637, pp. 276–295. Springer, Heidelberg (2007). https://doi.org/10.1007/978-3-540-74320-0_15
4. Garcia, S., Grill, M., Stiborek, J., Zunino, A.: An empirical comparison of botnet detection methods. computers & security 45, 100–123 (2014)
5. Gu, G., Zhang, J., Lee, W.: BotSniffer: Detecting botnet command and control channels in network traffic (2008)
6. Ja'fari, F., Mostafavi, S., Mizanian, K., Jafari, E.: An intelligent botnet blocking approach in software defined networks using honeypots. J. Ambient Intell. Humanized Comput. **12**(2), 2993–3016 (2020). https://doi.org/10.1007/s12652-020-02461-6
7. Laskov, P., et al.: Practical evasion of a learning-based classifier: a case study. In: 2014 IEEE Symposium on Security and Privacy, pp. 197–211. IEEE (2014)
8. Lee, S., Abdullah, A., Jhanjhi, N., Kok, S.: Classification of botnet attacks in IoT smart factory using honeypot combined with machine learning. PeerJ Comput. Sci. **7**, e350 (2021)
9. Lin, G., Wen, S., Han, Q.L., Zhang, J., Xiang, Y.: Software vulnerability detection using deep neural networks: a survey. Proc. IEEE **108**(10), 1825–1848 (2020)
10. Linhai, Y.: Research on distributed denial of service attack model and security defense strategy based on network layer. Cyberspace Secur. **11**(10), 9 (2020)
11. Liu, S., Dibaei, M., Tai, Y., Chen, C., Zhang, J., Xiang, Y.: Cyber vulnerability intelligence for Internet of Things binary. IEEE Trans. Ind. Inform. **16**(3), 2154–2163 (2020)
12. Miao, Y., Chen, C., Pan, L., Han, Q.L., Zhang, J., Xiang, Y.: Machine learning based cyber attacks targeting on controlled information: a survey. ACM Comput. Surv. **54**(7), 136:1–136:36 (2021)
13. Nagaraja, S., Mittal, P., Hong, C.Y., Caesar, M., Borisov, N.: Botgrep: finding p2p bots with structured graph analysis. In: USENIX Security Symposium, vol. 10, pp. 95–110 (2010)
14. Qiu, J., Zhang, J., Luo, W., Pan, L., Nepal, S., Xiang, Y.: A survey of android malware detection with deep neural models. ACM Comput. Surv. **53**(6), 1–36 (2020)
15. Wang, M., Zhu, T., Zhang, T., Zhang, J., Yu, S., Zhou, W.: Security and privacy in 6G networks: new areas and new challenges. Digit. Commun. Networks **6**(3), 281–291 (2020)
16. Zhang, J., Pan, L., Han, Q.L., Chen, C., Wen, S., Xiang, Y.: Deep learning based attack detection for cyber-physical system cybersecurity: a survey. IEEE/CAA J. Automatica Sinica (2021). https://doi.org/10.1109/JAS.2021.1004261
17. Zhou, J., Xu, Z., Rush, A.M., Yu, M.: Automating botnet detection with graph neural networks. arXiv preprint arXiv:2003.06344 (2020)

Vulnerability and Transaction Behavior Based Detection of Malicious Smart Contracts

Rachit Agarwal, Tanmay Thapliyal$^{(\boxtimes)}$, and Sandeep Kumar Shukla

CSE Department, IIT Kanpur, Kanpur, India
{rachitag,tanmayt,sandeeps}@iitk.ac.in

Abstract. Smart Contracts (SCs) in Ethereum can automate tasks and provide different functionalities to a user. Such automation is enabled by the 'Turing-complete' nature of the programming language (Solidity) in which SCs are written. This also opens up different vulnerabilities and bugs in SCs that malicious actors exploit to carry out malicious or illegal activities on the cryptocurrency platform. In this work, we study the correlation between malicious activities and the vulnerabilities present in SCs and find that some malicious activities are correlated with certain types of vulnerabilities. We then develop and study the feasibility of a scoring mechanism that corresponds to the severity of the vulnerabilities present in SCs to determine if it is a relevant feature to identify suspicious SCs. We analyze the utility of the severity score towards detection of suspicious SCs using unsupervised machine learning (ML) algorithms across different temporal granularities and identify behavioral changes. In our experiments with on-chain SCs, we were able to find a total of 1094 benign SCs across different granularities which behave similar to malicious SCs, with the inclusion of the smart contract vulnerability scores in the feature set.

Keywords: Blockchain · ML · Suspect identification

1 Introduction

Ethereum was the first blockchain platform to enable programming Turing-complete smart contracts (SCs). However, the use of SCs in such expressible language also opens the doors to vulnerabilities and bugs. These enabled various participants in the Ethereum platform (organizations and individuals) to exploit the vulnerabilities for malicious activities (such as Bitpoint Hacks). An Ethereum account is *malicious* if it performs, facilitates, or is suspected to be involved in different illegal activities such as *Phishing*, *Gambling*, and *Ponzi* schemes. While these malicious activities are often socially motivated (for example, Gambling and Phishing) and do not exploit and SC vulnerabilities, in several other cases (for example, Lendf Hack and Akropolis hack[1]), malicious activities are carried out due to the exploitation of bugs and vulnerabilities.

---

[1] Akropolis hack: https://zd.net/3wKtANQ.

© Springer Nature Switzerland AG 2022
W. Meng and M. Conti (Eds.): CSS 2021, LNCS 13172, pp. 79–96, 2022.
https://doi.org/10.1007/978-3-030-94029-4_6

In [3], the authors survey different vulnerabilities that exist in an SC. While bugs and vulnerabilities are the main focus of their survey, it lacks behavioral understanding as it does not consider the transactions performed by an SC. In Ethereum, there are two types of transactions performed by an SC: internal and external. External transactions are recorded on the ledger, while the internal transactions are not recorded on the ledger but can be obtained using Ethereum Virtual Machines (EVM). Internal transactions are mainly of 5 types: CALL, CALLCODE, SUICIDE, DELEGATECALL, and CREATE. Briefly, a CALL transaction refers to a transaction where an SC invokes another SC. In a CALLCODE and DELEGATECALL, the caller calls an SC on behalf of another SC. Note, the DELEGATECALL opcode is a newer version of the CALLCODE opcode. SUICIDE opcode allows an SC to self-destruct, causing all the SC's internal transactions to be lost. At the same time, a CREATE opcode enables an SC to create a new SC. Due to such types of opcodes, a malicious SC could create multiple SCs to evade detection. As identified in [8], irrespective of whether they are malicious or not, most of the created SCs have the same code and therefore have the same vulnerabilities. Nonetheless, only a fraction of SCs that have vulnerabilities are exploited [16]. SCs that have similar code to those exploited are not detected/marked as malicious as those SCs themselves are not exploited.

Thus, we ask if *we can train a machine learning (ML) algorithm to detect an SC that shows malicious behavior (even when the SC is not marked as malicious) and has vulnerabilities*. Note that when an SC is not marked malicious, we can only consider it as a *suspect* and cannot officially label it as malicious. Various state-of-the-art approaches exist that detect malicious and vulnerable SCs, such as [1,8]. In [1], the authors consider SCs as regular accounts (or the Externally Owned Accounts (EOA)s) and neglect SC vulnerabilities and their internal transactions. Further, as malicious activities can be of different types, the approach does not correlate vulnerabilities exploited by a particular malicious activity. In [8], the authors use various SC code analysis tools in the context of the Decentralized Application Security Project (DASP)[2] and identified the top 10 vulnerabilities. Nonetheless, more than 36 (still growing) vulnerabilities are identified under Smart contract Weakness Classification (SWC)[3], some of which are not considered in prior work.

Thus, we are motivated to answer: *(Q1)* is there a correlation between a particular malicious activity and a vulnerability in the SC, and if so, does the severity of a vulnerability correspond to its exploitability in committing malicious activities? *(Q2)* is the vulnerability severity score an important feature to learn by an ML algorithm that aims at detecting malicious accounts, and should it be used to identify malicious SCs (or the suspects)? And *(Q3)* do the SCs that are not marked malicious also show malicious behavior in different temporal granularities (observations in different temporal scales), and can the usage of severity score as a feature along with other temporal features detect such SCs?

---

[2] DSAP: https://dasp.co/.
[3] SWC Registry: https://swcregistry.io/.

To answer these questions, we first use different SC vulnerability analysis tools to analyze SCs. In parallel, we generate their transaction-based, graph-based, and temporal-based features from both internal and external transaction data available via Etherscan [9]. The list of malicious SCs is available using [10]. As this list is limited, we develop a CREATE transaction-based graph to identify all those SCs that a malicious SC creates, assuming that the parent and child of the malicious SC will be malicious. Of course, we understand that, in reality, this assumption might fail due to transaction behavior. Our analysis reveals that for a subset of SC vulnerabilities, such as those under *CWE-841*, there exists a correlation between them and the transaction behavior shown by SCs. Motivated by this, we include severity score as a feature and apply different ML algorithms to answer whether the inclusion of such feature improves the detection of malicious SCs. Towards this, we create two different datasets, one that includes both severity and transaction-based features and the other, which has only transaction-based features. We analyze them and check if better silhouette scores are obtained when compared with the silhouette scores when the severity score was not included in the feature set. Our analysis reveals that not all SCs with similar vulnerabilities cluster together. In the process, we also validate the findings of [16] and observe that the existence of a 'vulnerability' does not imply that an SC is exploited. Furthermore, we divide the dataset into different temporal granularities such as Daywise (1-Day), 3-Day, and month (1-Month). On the identified sub-datasets, we recompute the features and apply the unsupervised ML algorithm to identify the probability of a benign SC being malicious. We observe that SCs do show malicious behavior in different temporal granularities. For instance, when observing the behavior of SCs in the 3-Day granularity, we find 24 benign SCs that behave similar to the malicious SCs over time, and thus, we consider them to be suspects. We also observe a difference in the number of benign SCs that behave similar to malicious SCs in a particular granularity when we include the severity score as a feature. For instance, when we use both transaction and severity-based features in the Month-wise granularity, we discover 1066 suspects compared to 866 suspects when we consider only transaction-based features.

In summary, our core contributions are:

- We present a *mapping between different vulnerability vocabularies in the domain of SC vulnerabilities*. This mapping is based on existing vulnerabilities present in the deployed SCs. It provides a clear understanding of multiple names with which a particular vulnerability is referred to by the SC code analysis tools.
- We validate that *all the SCs with vulnerabilities are not usually exploited*, and the *severity scores of the vulnerabilities do not impact the transaction behavior*. Our findings are based on the currently known ground truth about the SCs. Nonetheless, *there exists a correlation between the type of malicious activity and the vulnerability*. For example, CWE-362 vulnerability is only present in the SCs related to Phishing schemes amongst the malicious class.

– Using our methodology, we *identify 2 SCs as potential suspects* using the *K-Means* algorithm. K-Means performed the best among the set of different unsupervised ML algorithms. Note that the behavior of SCs could change across different temporal granularities. An analysis across the temporal granularities reveals 892 SCs (866 in 1-Month + 24 in 3-Day + 2 in 1-Day) as potential suspects when only transaction-based features are used. This number changes to 1094 SCs (1066 in 1-Month + 24 in 3-Day + 4 in 1-Day) as potential suspects when both transaction-based features and severity score-based features are used.

In the rest of the paper, in Sect. 2, we present an overview of the state-of-the-art techniques used to detect accounts behaving maliciously in blockchains. In Sect. 3, we present a detailed description of our methodology. This is followed by an in-depth evaluation along with the results in Sect. 4. We finally conclude in Sect. 5 providing details on prospective future work.

2   Background and Related Work

Various studies focus on the detection of malicious activities in the blockchain. While some focus on detecting vulnerabilities in the SCs, others analyze the blockchain by observing the transaction-based features or using both transactions and source code-based features. Note that very few focus on studying the impact of existing vulnerabilities and transaction behavior in determining suspects. In the following subsections, we briefly survey the vulnerabilities detected by the tools and approaches that use transactions to classify or cluster malicious accounts.

2.1   Vulnerability Detection

Different vocabularies exist which classify various SC vulnerabilities. In [7], the authors use the NIST bug framework[4] to categorize different SC vulnerabilities into four categories: *Security, Operational, Functional,* and *Developmental.* Similarly, in [8], the authors map different SC vulnerabilities to the top 10 DASP identified vulnerabilities. Nonetheless, specific to vulnerabilities present in the SCs, the Smart contract Weakness Classification (SWC) vocabulary exists. Although generated from Common Weakness Enumeration (CWE: a broader classification nomenclature for vulnerabilities), SWC is yet to be standardized and, in some cases, does not cover all the vulnerabilities. For the sake of completeness, in Table 1, we present the relation between these vocabularies barring the NIST bug framework as there are pending updates to the framework. We find that as these vocabularies are not standard, the interpretation of a vulnerability and its severity varies. For severity scores, we obtain them from individual code analysis tools, and in case of a clash, we chose an interpretation that has a higher severity. Note that Table 1 lists only those vulnerabilities that are present in all the

---

[4] NIST Bug Framework: https://samate.nist.gov/BF/.

SCs in our dataset. Here we also note that *Bad Randomness* and *Front Running* vulnerabilities defined in DASP are not present in any SC. Furthermore, if a vulnerability is not present in some vocabulary, we mark the corresponding cell in the table with a "−". We put in our best effort to minimize the number of "−" and underline those we infer.

Table 1. Vulnerabilities.

Severity	Vulnerability	DASP-10 [8]	SWC	CWE	Severity	Vulnerability	DASP-10 [8]	SWC	CWE
H	Arbitrary-send†	Acc. Control	124	123	H	Uninitialized state†	Unknown	109	824
H	Ether send ‡		105	284	H	Uninitialized storage†		109	824
H	Unprotected self destruct† ‡ ◊		106	284	H	Shadowing state†		119	710
H	Delegate call†‡		112	829	H	Locked Ether◁†		-	-
H	tx-origin◁ ‡ †		115	477	M	Uninitialized local†		109	824
H	Integer Overflow‡◊ • ‡	Arith.	101	682	M	Constant function†		-	-
H	Integer Underflow‡◊•		101	682	M	Shadowing abstract†		119	710
M	Signedness bugs•		<u>101</u>	682	M	ERC20 returns false◁		<u>135</u>	1164
M	Truncation bugs•		<u>101</u>	682	M	Incorrect Blockhash◁		<u>104</u>	252
M	Callstack bug•◊	DoS	<u>113</u>	703	M	Balance Equality†◁		<u>132</u>	697
M	Overpowered role◁		-	-	L	Usage of Assembly†◁		-	<u>695</u>
M	Gas Limit in Loops◁		<u>128</u>	400	L	Pragmas version◁		102	937
M	Transfer in Loop◁†		<u>113</u>	703	L	Should not be view◁		-	-
L	Array Length Manipulation ◁		<u>128</u>	400	L	Bad Visibility◁		108	710
L	Multiple Calls‡		113	703	L	Shadowing-builtin†		119	710
H	Reentrancy-eth†‡#	Reentrancy	107	841	L	Shadowing-local†		119	710
M	Message call to ext. contract‡		107	841	L	Hardcoded address◁		-	<u>547</u>
M	Call without data◁		<u>107</u>	841	L	Deprec. Constructions◁		111	477
M	Reentrancy-no-eth†#		107	841	L	Extra gas in loops◁		<u>128</u>	400
L	Reentrancy-benign†#		107	841	L	Redun. fallback reject◁		<u>135</u>	1164
L	State change after ext. call‡		107	841	L	Revert require◁		<u>123</u>	573
H	Unchecked call return value◁‡	U.L.C	104	252	L	Exception State‡		<u>110</u>	670
M	Unused return†		<u>135</u>	1164	M	TOD‡		114	362
L	Send◁		<u>104</u>	252	M	Timestamp mani.◁ † ‡ • ◊		116	829

† = Slither, ‡ = Mythril, ◁ = SmartCheck, ◊ = Oyente, • = Osiris, ^{severity} H = high, M = medium, L = low, ⁻ inferred by us and not directly present in SWC and CWE, -: not present in the vocabulary, $^{Acc.}$ Access, Arith.: Arithmetic, U.L.C.: Unchecked lowlevel calls

Different state-of-the-art approaches use static, dynamic, taint analysis, and symbolic execution of the source code to detect vulnerabilities in SCs. In [8], the authors analyze the source codes of different SCs for vulnerabilities using nine different SC code analysis tools. Note that we do not survey the different SC code analysis tools as it is out of the scope of this work. However, we provide a brief description and our analysis of results in [8]. As initial results, for the data they had, they found only 42% of SC had verified unique source codes in which only 4.8% were unique. The analysis of these unique SCs reveals that the vulnerabilities under the categories such as *Access Control*, *Denial of Service*, and *Front Running*, present under the DASP-10 vocabulary, are not captured well by most of the tools. Further, they found that analysis tools such as Mythril [14] and Slither [12] together identify the maximum vulnerabilities present in the DASP vulnerability set. Individually, Mythril performs the best and identifies 27% of vulnerabilities present in their dataset. Note that slither uses a call graph to identify vulnerabilities. In another work, in [13], the authors compare different SC analysis tools such as Remix, Slither, SmartCheck, Oyente, Mythril, and Securify and determine that Slither performs the best as it detects at least one vulnerability from different vulnerability classes considered by them. In [4], the authors categorized the SC vulnerabilities into three groups: blockchain platform-based vulnerabilities (such as *Transaction Ordering Dependence (TOD)*, *Random Number*, *Timestamp*), EVM based (such as *CallStack Depth, Lost Ether*), and Solidity based (12 vulnerabilities including *Reentrancy, Unchecked Calls*, and *tx.origin*). They analyze these vulnerabilities using 27 different SC analysis tools. They concluded that Mythril could identify 75% of the blockchain-based vulnerabilities, while SmartCheck [18] detects 72% of all the vulnerabilities. In [15], the authors also reach a similar conclusion.

In [19], the authors study six types of vulnerabilities: *Integer Overflow and Underflow, Transaction-Ordering Dependence, Callstack Depth Attack, Timestamp Dependency*, and *Reentrancy Vulnerabilities* while not considering *Denial of Service (DoS)* and *tx-origin*. They develop an ML-based model called *ContractWard*, which at first uses 'SMOTE' for over-sampling data and then undersampling data points that have neighborhood relations. They then extract 1619 features using 2-gram analysis of opcodes of SCs to automatically detect the vulnerabilities mentioned above and then apply 'XGBoost'. Here, 2-gram refers to a set of 2 tokens, where the probability of occurrence of a token depends on the previous token. They achieve an average F1-score of 0.96 on their dataset. However, ContractWard has two limitations: *(i)* it uses 'SMOTE' to oversample dataset, and *(ii)* do not consider transactions carried out by SCs, which are important as not all vulnerable SCs are necessarily exploited [16].

2.2   Transaction Based Techniques

In [1], the authors use the transactions of accounts on the Ethereum blockchain to develop temporal transaction features to identify malicious accounts. They first survey different state-of-the-art algorithms used to detect malicious accounts on different permission-less blockchains. They identify that transactions on the Ethereum blockchain show bursty behavior for the degree, gas-price, inter-event time, and balance. Using burst-based features, the authors developed an ML pipeline that achieves high recall (>78%) in detecting the entire malicious class in their dataset. However, they considered SCs as equivalent to EOAs and did not consider internal transactions. In a follow-up study, in [2], the authors analyze different malicious activities and identify that neural networks perform best while detecting any adversarial attack that uses transaction behaviors as a feature vector.

In [17], the authors cluster EOAs and SCs in the Ethereum based on their transactions. They use a dataset containing transactions of 526121 accounts. They use the *birch* algorithm to perform hierarchical clustering on their dataset and only study the top 10 clusters with the maximum number of accounts. They observed that many malicious accounts cluster together.

In another approach, in [6], the authors use Graph Convolutional Networks (GCN) to detect EOA and SCs associated with Phishing activities. They first acquire the transactions of accounts marked as 'Fake Phishing' from Etherscan and build a graph with accounts as its nodes and the transactions carried out by them as edges. They obtain a graph with 13 connected components where they choose only the largest connected component subgraph for their analysis. They use features such as 'Indegree', 'Outdegree', and 'Number of Neighbors' to detect Phishing accounts. On the feature vector, GCN achieves an average F1 score of 0.24. Although they consider both internal and external transactions, they assume that all the accounts which are related to or carry out internal transactions are 'non-phishing' EOAs and SCs.

In [11], the authors use XGBoost to detect illicit EOAs and SCs (including tokens such as ERC-20) in the Ethereum blockchain. They use only 2179 malicious accounts and 2502 normal accounts for their experiments and use 42 transaction-based features such as *Time_diff_between_first_and_last*, *min_value_received* and *min_value_sent*. They also rank features based on their importance and find *Time_diff_between_first_and_last* as the most important feature while *ERC20_ Most_Sent_Token_Type* being the least important one. Using these features, they achieve an average accuracy of 0.963. However, they use a highly under-sampled dataset, with a ratio of 1:1.14 between the malicious and benign classes. This does not represent the actual distribution of malicious and benign accounts in Ethereum, which has more than 14 million unique accounts.

Note that the above state-of-the-art approaches do not consider source code-based features to carry out a behavioral analysis on the SC in Ethereum. To address such an issue, in [5], the authors use both transaction and source code-based features to detect honeypot accounts in the Ethereum blockchain. They use a dataset with 16163 accounts, of which 295 are marked as honeypots by HoneyBadger's repository[5]. Using transaction-based features (such as *Transaction Count* and *Transaction Value*) and source code-based features (such as *hasByteCode* and *hasSourceCode*), the authors train 'XGBoost'. They achieve an 'Area Under the Receiver Operating Characteristics' mean score of 0.968 on their dataset. They, however, do not consider the temporal aspects of blockchain transactions.

Moreover, in [5], the authors use features extracted from the opcodes of an SC and their transactions. However, they do not consider the vulnerabilities that are present and are exploited by attackers as features. To the best of our knowledge, ours is the first work in the field of blockchain security that considers both temporal behavior (extracted using both internal and external transactions) and vulnerabilities present in an SC to detect potential suspects.

3 Methodology

In this section, we describe our approach in detail. We first obtain the source codes of all the SCs available in the Ethereum blockchain using the Etherscan

---

[5] HoneyBadger: https://github.com/christoftorres/HoneyBadger.

APIs [9] and their malicious tags (using the Etherscan label cloud service [10]) and the internal and external transaction data.

With CREATE-type internal transactions, a malicious SC can create several child SCs. These child SCs further develop several other SCs. Most child SCs have the same source code as their malicious parent SC; therefore, their vulnerabilities are also the same. In general, despite not showing malicious transaction behavior, such child SCs could also be exploited and thus should be marked as suspects. On the other hand, a parent of the malicious SC could be unaware of the vulnerabilities or could have written the malicious SC with specific malicious intent. Therefore, parents of malicious SC should also be marked as suspects. This leads to marking the entire parent-child chain of the malicious SCs as malicious. Contrary to this, in specific cases, valid organizations create numerous user-centric SCs where only a particular SC is involved in performing a malicious activity (for example, Bittrex). In such cases, marking the entire chain suspicious would be incorrect. Thus, where we know that an organization developed a particular SC for a specific purpose, we do not create its chain. While for other malicious SCs we do. Currently, in state-of-the-art approaches, such SCs are not considered while training ML algorithms. We thus include such SCs as malicious SCs in our study.

Next, we identify code similarity between all the suspect SC identified using the above assumption. Here, we use the hashing technique as used in [8], where we generate a hash of the source code and identify unique hashes. Note that if two or more SCs have different source codes, they are lexicographically different, resulting in SCs having different hashes. We do not consider OPCODEs to generate the hashes. Hashes reduce the computational resources needed to identify vulnerabilities in the SCs as SCs with identical hashes will have the same vulnerabilities. We analyze the uniquely identified SCs with five different vulnerability detection tools: SmartCheck, Mythril, Oyente, Slither, and Osiris. While Slither and SmartCheck carry out static analysis on SC source code, Mythril relies on taint analysis. Oyente uses symbolic execution of the SC source code to detect vulnerabilities, while Osiris extends this functionality to detect Integer related bugs. Here, we choose these tools as they collectively detect most of the vulnerabilities described in Table 1.

Along with the ground truth available about the type of malicious activity a particular SC is involved in and all the identified SC vulnerabilities, we then study the correlation between the vulnerabilities and the type of malicious activity they associate with. In the process, we also study if exploitability is reflected due to the existence of any vulnerability. A severity score is associated with all the vulnerabilities present in the SCs. Also, an SC can have multiple vulnerabilities with different severity. Let V^i be the set of all vulnerabilities present in an SC i and let a vulnerability $j \in V^i$ have a severity score S_j. A *severity score* (Ss_i) for an SC i is defined as Eq. 1. Note that the Ss^i represents the average severity score.

$$Ss^i = \frac{\sum_{\forall j \in V^i} S_j}{|V^i|} \tag{1}$$

Besides computing the severity score, we analyze the transaction dataset and identify different behavioral features. These features are based on the approach defined in [1]. These features capture *(i)* temporal behavior along with static properties, and *(ii)* provide the best results in terms of recall on malicious class. For our study, we create two different data configurations using such features (one with the severity score defined above and another without). In this work, we study different unsupervised ML algorithms (tested on different hyperparameters) such as K-Means (n_clusters $\in [3, 26]$), HDBSCAN (min_cluster_size $\in [2, 1735]$), Spectral clustering (n_clusters $\in [3, 26]$), Agglomerative clustering (n_clusters $\in [3, 26]$), and OneClassSVM (Degree $\in [2, 10]$, kernel='poly') to identify the algorithm (and their hyperparameters) that perform the best based on silhouette score. The choice of hyperparameter values reflects the computational power available to us. Ideally, to establish correlation, SC's should cluster better when we use the severity score as a feature along with the transaction and temporal-based features.

We analyze the datasets based on different temporal granularities to understand behavioral aspects and whether SCs show persistent malicious behavior over time. In a particular temporal granularity, we consider only those SCs and their transactions that occur in a specific period defined by the temporal granularity. For instance, in a 1-Day temporal granularity, we consider only the transactions in a given 1-Day period. Note that there could be multiple periods in a particular granularity. We study the behavior across four different temporal granularities in this work: 1-Day, 3-Day, 1-Month, and aggregated (ALL). Henceforth, whenever we use the word 'segment', we refer to a particular period from a granularity as mentioned above. We create two different feature vectors for each segment: one that contains both severity score and transaction-based features and another that only has transaction-based features. For each segment and the two sets of feature vectors, we use the unsupervised algorithm that performs best (identified via Q2) with an assumption that the same algorithm would serve the best across the different segments. For each segment in each granularity, we then determine the largest cluster with a maximum number of malicious SCs. We then compute the cosine similarity amongst benign and malicious SCs present in that cluster to identify which benign/unmarked SCs behave similarly to the malicious SCs. Our motivation for using the largest cluster only is: *(a)* we assume that all malicious SCs show similarity and cluster together, and *(b)* choosing such cluster reduces the search space and is within the computation limits available to us. We acknowledge that there exist several other metrics to identify similarity scores (such as *Jaccard*), and using such metrics will give different results. But we use the cosine similarity metric because *(i)* it is more popular and widely adopted, and *(ii)* was used in [1] (we use their features) to detect the suspects. A very high cosine similarity score ($CS_{ij} \rightarrow 1$) between a malicious SC, i, and a benign SC, j, indicates that j's behavior is suspicious in that segment. For each granularity, as j could change its behavior over time, we associate j with a probability ($p(j)$) of being malicious. Over all the segments, a high probability ($p(j) = 1$) means that the SC should be marked malicious considering that temporal granularity.

To identify this probability, we use the same method as in [1]. We then compare across different granularities to determine which suspect SCs are common to say that the used ·granularity does not impact their behavior.

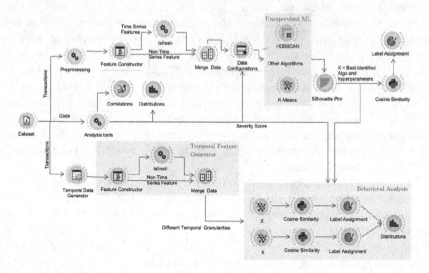

Fig. 1. ML pipeline.

In summary, Fig. 1 presents the entire pipeline. We first use SC analysis tools to do a code-based analysis and detect vulnerabilities in the source code of SCs present in our dataset. We then correlate the detected vulnerabilities present in SCs with different malicious activities. Since each vulnerability has an associated severity, we assign a severity score to each SC. This score is averaged over all the vulnerabilities present in that SC. To determine the usefulness of the severity score towards detecting malicious SCs, we first compute different transaction and graph-based temporal features. We then create different datasets, one with both transaction and severity-based features and the other with only transaction-based features. We then use different unsupervised ML algorithms on the above-created datasets across different temporal granularities (such as 1-Day, 3-Day, and 1-Month) and analyze the results to determine the usefulness of the severity score as a feature.

4   Evaluation and Results

This section provides an in-depth analysis of our approach towards answering Q1, Q2, and Q3 and presents our results. All our analysis is performed using Python version 3 and its associated libraries such as scikit-learn.

4.1   Data

We use the Etherscan blockchain explorer APIs [10] to acquire SCs which are associated with malicious activities. This results in a list of 403 SCs marked malicious until 28th August 2020 (block number 10747845) since the induction of Ethereum. Note that the tag (including malicious ones) assigned to an SC in Etherscan is crowd-sourced, i.e., any person can suggest the tag. Since we cannot validate the correctness of these tags, we assume that SCs are correctly associated with different malicious activities. Although there are multiple malicious tags present via Etherscan, such as those described in [2], until the data collection time, malicious SCs are only associated with four malicious activities: Phishing, Gambling, High-Risk, and Ponzi Scheme.

For all these 403 SCs, using their internal transactions and the heuristics described in Sect. 3, we identify ≈2 Million SCs that are either successor or predecessor to a malicious SC. For all the marked malicious SCs, we observe that:

- 377 out of 403 marked SCs, are created by EOAs. These SCs do not have any CREATE type transactions and thus do not create any successor SCs.
- Out of the remaining 26 marked malicious SCs, only 8 SCs (that also have EOAs as their parents) create a total of 52 SCs. However, these 52 SCs did not create any new SCs.
- Out of the remaining 18 marked malicious SCs, these SCs have 12 unique SCs as their parent. Although these 18 SCs did not create any successor SCs, their 12 parents created many SCs.

Out of all the SC, we observe that only 46 unique hashes exist and corresponding to 46 unique SC codes. Note that this number represents SCs for which source codes are available. There are 165 malicious SCs for which the source code is not available. We do not consider them in our study because the source code could be different and our feature vector depends on the severity score obtained using vulnerabilities present in the SCs. In all 46 unique hashes identified, 38 unique hashes belong to 38 marked SCs. This also means there are only 38 unique codes present between the remaining 238 marked SCs. In the remaining 8 SCs that are unmarked and detected using our graph analysis, 7 SCs lie in the graph's component created using seven different marked Phishing SCs, and 1 SC lies in the graph component created using 1 marked Ponzi scheme-based SC. As we have limited computational resources, analyzing both internal and external transactions of all these 2 million SCs is practically not feasible for us. Thus, we restrict our analysis and consider a union of these 46 SCs and 47398 unique SCs identified by [8].

We identify 314614 vulnerabilities in total. Out of these, 314302 are present in the benign SCs, and 312 are present in malicious SCs across different severities (54 for high, 92 for medium, and 166 for low). These 312 vulnerabilities in the 46 malicious SCs are distributed as follows: 192 vulnerabilities are present in the Phishing based SCs, 95 vulnerabilities are present in the Gambling based

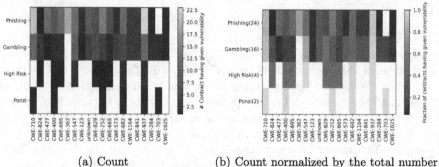

(a) Count

(b) Count normalized by the total number of SCs of that malicious type. The number in "()" on the y-axis represents the number of malicious SCs

Fig. 2. Distribution of vulnerabilities in every type of malicious activities present in the dataset.

SCs, 19 vulnerabilities are present in the High-Risk based SCs, and 6 vulnerabilities are present in the Ponzi scheme-based SCs. In Phishing SCs, there are 26 vulnerabilities with high severity, 62 with medium, and 104 with low severity. Gambling SCs have 21 vulnerabilities with high severity, 20 with medium, and 54 with low severity. High-Risk SCs have 6 vulnerabilities with high severity, 7 with medium, and 6 with low severity. Similarly, Ponzi scheme-based SCs have 1 vulnerability with high severity, 3 with medium, and 2 with low severity.

In Ethereum, on average, their are ≈6000 blocks created each day. Using such information, we develop segments for different temporal granularities. For 1-Day granularity, from the genesis block until our collection date, we have 1791 segments. Here each segment corresponds to 6000 blocks. For example, segment 1 contains transactions of considered SCs from genesis block until block number 6000. Similarly, for the segments in the 3-Day granularity, we consider transactions of considered SCs in 6000 × 3 blocks, and for segments in 1 month, we consider 6000 × 30 blocks. Thus we have 598 segments and 60 segments for the 3-Day granularity and the Month granularity, respectively.

4.2 Results

Our results pertain to the three research questions. Thus this section is divided into 3 parts, with each part referring to the research question defined in Sect. 1.

Q1: Correlation between malicious activities and Vulnerabilities and whether severity of a vulnerability correspond to its exploitability. One way to identify the correlation between malicious activities and vulnerability is to study the distribution of the vulnerabilities in the malicious activities associated with the SCs. Thus, we identify the fraction of malicious contracts related to a specific CWE vulnerability for each category of malicious activity.

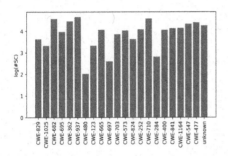
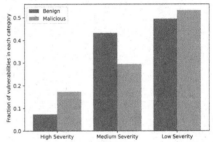

(a) Number of vulnerabilities of specific type across all SCs on a semi-log scale

(b) Fraction of vulnerabilities of a particular severity in SCs

Fig. 3. Distribution of vulnerabilities.

To understand the correlation, we normalize this and then study the relation, if any. Figure 2 shows both the number (cf. Fig. 2a) and the normalized count of a specific vulnerability (cf. Fig. 2b). From Fig. 2, we identify that none of the SCs related to Phishing type of malicious activity have CWE-703 vulnerability. However, vulnerabilities such as CWE-362, CWE-937, CWE-252, and CWE-710 are present in large numbers, with CWE-937 present in almost all malicious SCs. Here, we also note that CWE-362 (a medium severity vulnerability) is only present in the SCs related to the Phishing type malicious activity. Moreover, SCs tagged under malicious activities such as High-Risk, Gambling, and Ponzi labels do not report SC vulnerabilities under CWE-362. Further, we observe that SCs under the Ponzi scheme have 6 vulnerabilities: CWE-710, CWE-400, CWE-252, CWE-682, CWE-937, and CWE-703. We also observe that although each malicious activity has SCs with high severity vulnerabilities, such as CWE-841 and CWE-123, their frequency is less. From the above observations, we infer that in SCs corresponding to malicious activities, the vulnerabilities with high severity are less in number. However, just from the above observations we cannot say that, for example, if an SC has vulnerabilities related to CWE-362, it is involved in Phishing activity. Our inference is based on the fact that CWE-362 is also present in benign SCs. Nonetheless, it is possible that such benign SCs are, in reality, related to Phishing activity but are not marked as Phishing SCs.

Further, to check if benign SCs also have vulnerabilities, we plot the distribution of vulnerabilities. Figure 3a shows the distribution of the vulnerabilities across all the SCs on a semi-log scale. Here, we observe that vulnerability CWE-937 is most frequent and occurs in most SCs (including benign SCs), while vulnerability CWE-480 is least common. With respect to the severity score (cf. Fig. 3b), we observe that high severity vulnerabilities (23214 vulnerabilities in total) are also present in the benign SCs, but their fraction (= (number of high severity)/(total vulnerabilities in the considered class)) is less than those present in the malicious SCs. We observe similar behavior for low severity vulnerabilities. However, in this case, the fraction is much higher. On the other hand, the fraction is much higher for the benign class for the medium severity vulnerabilities.

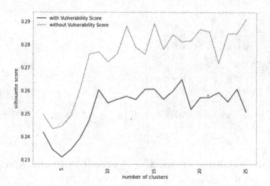

Fig. 4. Comparison between Silhouette scores.

Upon further investigation, we find a negligible difference between the average severity score of benign SCs (= 2.25) and malicious SCs (= 2.21). Further, as the difference between the fraction for malicious and benign class for each severity category is very small, we cannot say that severity of a vulnerability relates to exploitation.

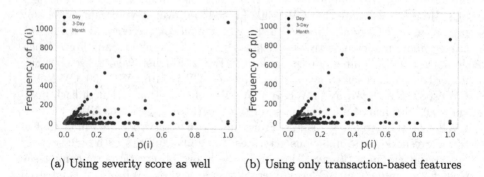

(a) Using severity score as well   (b) Using only transaction-based features

Fig. 5. Frequency plot of p(i) at different temporal granularities.

Q2: Importance of Severity score. To identify the importance of the severity score, we test the results obtained using different unsupervised ML algorithms and different data configurations. We find that K-Means performs best as it achieves the highest silhouette score amongst the various unsupervised algorithms we tested. For K-Means, when using both transaction and severity score-based features, for $K = 18$, the best silhouette score (0.26, while for other values of K, the silhouette score $\in [0.23, 0.26]$) is achieved. Similarly, when considering only transaction-based features, for $K = 25$, the best silhouette score (0.29, while for other K silhouette score $\in [0.24, 0.29]$) is achieved. Other unsupervised ML algorithms achieve lesser silhouette scores than K-Means. For HDB-SCAN and the tested hyperparameter configurations, we obtain silhouette scores

$\in [-0.11, 0.14]$. Here, for most of the hyperparameter configurations, the silhouette score was negative. For Spectral Clustering, the silhouette scores obtained were $\in [0.15, 0.22]$. While for the Agglomerative Clustering, we obtain silhouette scores $\in [0.12, 0.23]$. Finally for the OneClassSVM, silhouette score ranges between $[0.02, 0.09]$. From this analysis, we observe that K-Means provides better silhouette scores. Therefore, we use K-Means for our further analysis.

In the case of K-Means, Fig. 4 shows the silhouette scores for different K for the two different data configurations. From the figure, we observe that the silhouette scores obtained using severity score and transaction-based features always remain less than the scores obtained on using only transaction-based features. Thus, we infer that when such severity scores are considered a feature vector, the data is either more uniformly distributed or more densely distributed in a small feature space, causing overlapping clusters. The clusters thus formed are indistinguishable from each other, which in turn reduces the silhouette score. This also means severity score for inter-cluster analysis is not a good feature. However, severity score for intra-cluster analysis is a good feature to include as it is able to detect more SCs that show high similarity along the feature space with the malicious SCs.

Nonetheless, we calculate the similarity amongst benign and malicious SCs in the cluster with maximum malicious SCs. We find that the maximum similarity score between malicious and benign SCs is 0.74 when we consider both severity score and transaction-based features. This similarity score reduces to 0.73 when we consider only transaction-based features. The difference in these scores indicates that upon considering severity scores as features, the SCs have less distance between them in the feature space, i.e., are more closer. As the maximum is $\in \{0.73, 0.74\}$, we consider this to identify the benign SCs that are suspects and are within $\epsilon = 10^{-7}$. We find that there are 2 such benign SCs in both the cases. To further analyze the behavioral changes in the SCs, we identify the probability of an SC being a suspect. Note that in the ALL granularity, probability computation does not make sense as there is no notion of behavior change.

Q3: Understanding behavioral changes over time. Using the best unsupervised ML algorithm identified (K-Means algorithm) and the different data segments created using different temporal granularities, we investigate the cluster with the maximum number of malicious SCs. For temporal granularities other than the ALL granularity, we calculate the probabilities of benign SCs being malicious. Towards this, we run the K-Means clustering algorithm across all the temporal granularities and select the K (number of clusters) for which the maximum silhouette score is obtained for our analysis. Again, we investigate the cluster where the maximum number of malicious SCs are present for different temporal granularities. From the selected cluster, we then select those benign SCs as suspects where the $CS \to 1$ with malicious SCs, i.e., lie $\in 1 - \epsilon$ where $\epsilon = 10^{-7}$. Here, we find that:

- 1066, 24, and 4 SCs are identified as suspects and have $p = 1$ in 1-Month, 3-Day, and 1-Day granularity, respectively, when both transaction and Severity scores are used as features. Here, we do not identify any suspect SC that appeared across different temporal granularities.
- 866, 24, and 2 SCs are identified as suspects and have $p = 1$ in 1-Month, 3-Day, and 1-Day granularity, respectively, when only transaction-based features are used. Here as well, we do not find any suspect SC which is common across different temporal granularities.
- In these identified suspect SCs for the above two cases (when using severity score along with temporal features and when only using temporal features), we again do not find any common suspect SC.

Note that the difference in numbers of SCs identified as malicious for the two cases (with and without using severity score) is due to the reasons described in previous sub-section. That is, the data points become well-clustered when both severity score and transaction-based features are used, thereby increasing the intra-cluster density where the similarity score between malicious and benign SCs is high.

Figure 5 shows the distribution of the frequency of SCs with a particular probability across different temporal granularities with and without considering the severity scores as feature vectors. Here, we note that the distribution of the frequency of the SCs with a certain probability in 1-Day and 3-Day granularities are similar. This is because the difference between the timeframe represented by these granularities is less. Similarly, the distribution of the frequency of the SCs with a certain probability in the Daywise (1-Day, 3-Day) and 1-Month granularity is different. Again this is because the difference between the timeframe represented by these granularities is high. Also, these suspect SCs (that have a $p = 1$) carry out only a few transactions. From these figures, we infer that:

- Most of the SCs have a low probability score. This is represented by the overcrowding of the frequency of the SCs that have low probability scores (cf. Fig. 5). Note that the low probability score for an SC does not mean that the SC is not malicious. The probability score is less as the number of segments in which the SC was identified as suspects or actually did malicious transaction was less than the number of segments in which the SC carried out the other transactions.
- We observe that there are no common suspect SCs between different granularities. Therefore, we infer that the behavior of SCs is changing across different considered temporal granularities.

5   Conclusion and Discussion

The introduction of SCs has opened numerous possibilities for cyber-criminals to steal cryptocurrency and perform illegal activities. Many state-of-the-art approaches leverage ML-based techniques and study transaction behavior to detect accounts held by cyber-criminals. However, these techniques have limitations as they do not distinguish between SCs and other types of accounts. Furthermore, as SCs are programs targeting specific purposes, they have vulnerabilities.

In this work, we study the correlation between different malicious activities and the vulnerabilities present in SCs. We find that our results are consistent with those of [16] as we also do not observe any significant correlation between malicious activities and vulnerabilities. In the process, we also demonstrate the feasibility of using the severity scores of different vulnerabilities as a feature and detect possible suspects amongst the benign SCs. We find that the performance in terms of silhouette score is reduced when we use both the severity and temporal transaction-based features. The severity score feature seems to be a feasible feature for the problem at hand. We also detect different benign suspects across different granularities, such as 1-Day, 3-Day, and 1-Month, using the considered features to understand the temporal behavior changes. Here, we do not get any common suspects across different temporal granularities. This also indicates that the behavior of SCs changes across different considered temporal granularities.

Note that due to computational restraints, we only considered SCs with unique source codes for our analysis. With more computational resources, it is possible to consider all SCs and their transactions in the study. This could cause changes in the results where some suspicious SCs might occur throughout different temporal granularities. Nonetheless, it may also happen that some SCs have the same vulnerabilities, but their transaction behavior is different. Such aspects would lead to one SC being labeled as malicious while another being benign, which means that the transaction behavior is a more critical factor in identifying malicious SCs than vulnerabilities in the SCs.

Acknowledgement. This work is partially funded by the National Blockchain Project at IIT Kanpur sponsored by the National Cyber Security Coordinator's office of the Government of India and partially by the C3i Center funding from the Science and Engineering Research Board of the Government of India. We also thank authors of [8] for providing us their dataset which was partially used in our work.

References

1. Agarwal, R., Barve, S., Shukla, S.: Detecting malicious accounts in permissionless blockchains using temporal graph properties. Appl. Network Sci. **6**(9), 1–30 (2021)
2. Agarwal, R., Thapliyal, T., Shukla, S.: Detecting malicious accounts showing adversarial behavior in permissionless blockchains, pp. 1–15 (2021)
3. Alkhalifah, A., Ng, A., Kayes, A., Chowdhury, J., Alazab, M., Watters, P.: A taxonomy of blockchain threats and vulnerabilities. In: Maleh, Y., Shojafar, M., Alazab, M., Romdhani, I. (eds.) Blockchain for Cybersecurity and Privacy: Architectures, Challenges, and Applications, pp. 1–26. Taylor and Francis Group (2020)
4. Angelo, M., Salzer, G.: A survey of tools for analyzing ethereum smart contracts. In: International Conference on Decentralized Applications and Infrastructures, pp. 69–78. IEEE, Newark, CA (2019)

5. Camino, R., Torres, F., Baden, M., State, R.: A data science approach for detecting honeypots in ethereum. In: International Conference on Blockchain and Cryptocurrency (ICBC), pp. 1–9. IEEE, Toronto, Canada (2020)
6. Chen, L., Peng, J., Liu, Y., Li, J., Xie, F., Zheng, Z.: Phishing scams detection in ethereum transaction network. Trans. Internet Technol. **21**(1), 1–16 (2020)
7. Dingman, W., et al.: Defects and vulnerabilities in smart contracts, a classification using the NIST bugs framework. Int. J. Networked Distrib. Comput. **7**, 121–132 (2019)
8. Durieux, T., Ferreira, J., Abreu, R., Cruz, P.: Empirical review of automated analysis tools on 47,587 ethereum smart contracts. In: 42nd International Conference on Software Engineering, pp. 530–541. ACM/IEEE, Seoul, South Korea (2020)
9. Etherscan: Ethereum Developer APIs, October 2020. https://etherscan.io/apis, Accessed 09 Oct 2020
10. Etherscan: Label Word Cloud, October 2020. https://etherscan.io/labelcloud/, Accessed 09 Oct 2020
11. Farrugia, S., Ellul, J., Azzopardi, G.: Detection of illicit accounts over the ethereum blockchain. Expert Syst. Appl. **150**, 113318 (2020)
12. Feist, J., Greico, G., Groce, A.: Slither: a static analysis framework for smart contracts. In: 2nd International Workshop on Emerging Trends in Software Engineering for Blockchain, pp. 8–15. IEEE, Montreal, Canada (2019)
13. Gupta, B.C., Kumar, N., Handa, A., Shukla, S.K.: An insecurity study of ethereum smart contracts. In: Batina, L., Picek, S., Mondal, M. (eds.) SPACE 2020. LNCS, vol. 12586, pp. 188–207. Springer, Cham (2020). https://doi.org/10.1007/978-3-030-66626-2_10
14. Mueller, B.: Smashing ethereum smart contracts for fun and real profit. In: 9th Annual HITB Security Conference (HITBSecConf), pp. 1–54. HITB, Amsterdam, The Netherlands (2018)
15. Parizi, R., Dehghantanha, A., Choo, R., Singh, A.: Empirical vulnerability analysis of automated smart contracts security testing on blockchains. In: 28th Annual International Conference on Computer Science and Software Engineering, pp. 103–113. ACM, Markham, Canada (2018)
16. Perez, D., Livshits, B.: Smart contract vulnerabilities: vulnerable does not imply exploited. In: 30th USENIX Security Symposium, pp. 1–17. USENIX Association, Vancouver, B.C. (2021)
17. Sun, H., Ruan, N., Liu, H.: Ethereum analysis via node clustering. In: Liu, J.K., Huang, X. (eds.) NSS 2019. LNCS, vol. 11928, pp. 114–129. Springer, Cham (2019). https://doi.org/10.1007/978-3-030-36938-5_7
18. Tikhomirov, S., Voskresenskaya, E., Ivanitskiy, I., Takhaviev, R., Marchenko, E., Alexandrov, Y.: Smartcheck: static analysis of ethereum smart contracts. In: 1st International Workshop on Emerging Trends in Software Engineering for Blockchain, pp. 9–16. ACM, Gothenburg, Sweden (2018)
19. Wang, W., Song, J., Xu, G., Li, Y., Wang, H., Su, C.: Contractward: automated vulnerability detection models for ethereum smart contracts. Trans. Network Sci. Eng. **8**(2), 1133–1144 (2020)

A Novel Method of Template Protection and Two-Factor Authentication Protocol Based on Biometric and PUF

Hui Zhang[1,2], Weixin Bian[1,2(✉)], Biao Jie[1,2], and Shuwan Sun[1,2]

[1] School of Computer and Information, Anhui Normal University, Wuhu 241002, China
bwx2353@ahnu.edu.cn
[2] Anhui Province Key Laboratory of Network and Information Security, Wuhu 241002, China

Abstract. With the development of Internet technology and the change of network environment, it is particularly important to ensure the security and privacy of biometrics in the process of biometrics authentication. In this regard, we propose a novel identity authentication protocol based on cancelable biometric and Physical Unclonable Function (PUF) which uses the properties of PUF to generate the cancelable biometric and adds it to the complete authentication protocol, so as to realize the two-way authentication between the user and the server. Our authentication protocol makes full use of the characteristics of what users bring in and who users are, overcomes the shortcomings of the traditional key-based protocol, and connecting with the supervised learning algorithm SVM and elliptic curve Pedersen commitment, construct an effective, unique and cancelable biometric identity to replace original biometrics, thus improving the security and privacy protection of the biometrics template. At the same time, we analyze the accuracy of classification algorithms, the revocability and unlinkability of templates through experiments, which further ensures the security and legitimacy of the authentication protocol.

Keywords: Authentication · Biometric identification · Privacy protection · Cancelable biometrics · Physical unclonable function

1 Introduction

At present, biometric identification technology plays an important role in identity authentication system. Compared with traditional cryptographic authentication systems, biometrics have obvious advantages [1]. However, every coin has two sides. Using biological information for identity authentication in multi-server environment greatly increases the risk of biological information being stolen or lost. Hence, it is very important to generate cancelable biometrics as a research direction of biometrics protection scheme by avoiding the disadvantages of biometrics, improving the security of biometrics without losing the unique advantages of biometrics in authentication. The concept of cancelable biometrics was first proposed in 2007. In recent years, good progress has been made in

© Springer Nature Switzerland AG 2022
W. Meng and M. Conti (Eds.): CSS 2021, LNCS 13172, pp. 97–106, 2022.
https://doi.org/10.1007/978-3-030-94029-4_7

this direction, and many scholars have proposed various methods to generate cancelable biometrics in different research fields.

In order to obtain more secure biometrics, a protection policy using PUF for mutual authentication was recently proposed in the Internet of Things system, which can also be used in multi-service authentication system to protect biometrics [2–5]. PUF is a hardware integrated circuit that ensures the physical safety of the device. Any change in the input excitation is sufficient to result in a different response output. Even if the input is slightly different, the output will vary widely, and the characteristics of irreversibility, recognizability and revocability can be realized at the same time. This is very much in line with the characteristics that cancelable biometrics need to satisfy, so it makes sense to consider the integration of PUF and cancelable biometrics to protect the biometrics template.

By using PUF, the researchers achieved both the protection of the biological template and mutual authentication in a multi-service environment. In 2016, Arjona et al. [6] first introduced a two-factor access control system based on device and user internal identifiers where device identifiers produced by PUFs are XORed with binary, fixed-length fingerprint data to get biometric templates. By 2018, Arjona [7] proposed a stronger template protection strategy by fully integrating PUF data with detailed data of all fingerprints, but the underlying features based on biometrics in feature extraction hindered the realization of better recognition performance to some extent. Based on the most popular deep learning algorithm at present, Liu Y et al. [8] used random projection and deep belief network to process biological data, and combined it with password used for user authentication, giving full play to the advantages of biological password system and feature conversion method. Yang W et al. [9] also proposed a novel template protection algorithm. Firstly, the binary decision diagram (BDD) was used to generate an irreversible transformation template of the original template, and then the multi-layer extreme learning machine (ML-ELM) was used for training. It is worth noting that the above schemes are revocable changes based on the original features. In contrast, no template protection technology can perfectly meet all the biological characteristics and application requirements. Due to the complex and changeable test environment, it is not universal and does not combine with the identity of the equipment, so there are risks in security.

To address these problems, in this paper, we propose a safe and effective method for generating cancelable biometrics using PUF that can be extended to other biometric features and more classification algorithms. Specifically, our scheme is innovative in classification results, unlike existing schemes which generate revocable biometrics based on original features. The new combination of biological information and device identification in proposed scheme has realized both biometric template protection and two-factor authentication of user and device. The whole part of protocol and experimental analysis based on ORL public face data set demonstrates the effectiveness of our proposed method.

2 Method

2.1 System Model

Our system model consists of three entities, as shown in Fig. 1.

(1) Endpoint UD: A terminal connecting the user to the device. The user uses the face for identity authentication, and the device is embedded with an image sensor and a PUF.
(2) IDP: Identity Provider, a trusted agency for encrypting biometrics in a multi-server authentication system and for use by third-party servers for authentication.
(3) Server: Service provider, the entity that authenticates with the user before legitimately using the service provided by the server.

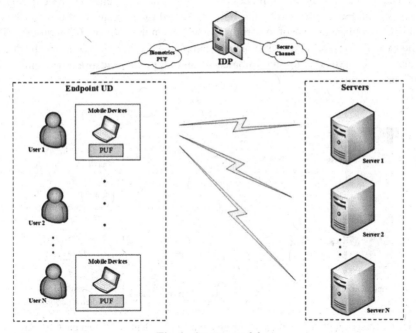

Fig. 1. System model

2.2 Method Description

Generation of Cancelable Biometric Identity. In order to avoid the drawbacks of biometrics being linked to the user for life, we use a novel approach that combines the user's class label L with the device's response R_i to generate a cancelable biometric identity BID instead of the direct use of the original biometrics. This BID is unique and safe. When the system encounters a malicious attack, it can be cancelled, which will not reveal the user's real biological information, but also generate a new BID based on the same biometric. In particular, we hide the BID perfectly in the Pedersen promise based on elliptic curve, register the identity token IDT with IDP digital signature, and when applying for retraction, IDP will issue a new IDT based on different class labels and classifiers to associate with the user. Figure 2 clearly shows the details of how we generated the BID based on facial features and SVM classification which can extend to other biometric authentications and classification algorithms. In this process, we first randomly assign a unique class label to each registered user in the authentication system and compose training data based on the face images of all registered users. Then, feature extraction based on eigen faces is carried out on the expanded training data at IDP, and the multi-classification SVM algorithm of machine learning is used to train the classifier. The obtained classifier is encrypted and stored in the client, which is used to output the correct class label L of legitimate users' face features during authentication. Subsequently, all the data related to the original biometric are revoked to improve the protection of user privacy. PUF, on the other hand, acts like a fingerprint on a device, uniquely identifying the device in the authentication process. The response R_i is generated by entering the challenge randomly generated by the IDP and linked to the result L of the classifier to get a BID to replace the use of the original biometric in the authentication process. This method not only protects the user biometric template from the perspective of the device, but also can be extended to other biometrics and more classification algorithms, with higher universality.

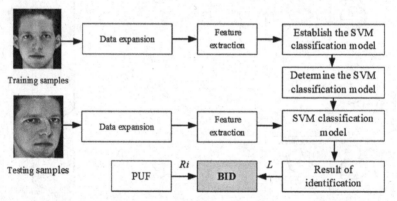

Fig. 2. The generation process of cancelable biometrics identity BID on the ORL dataset

Protocol Description. This section specifically describes the user authentication protocol and key agreement based on multi-server environment. We applied the cancelable biometrics identity BID into the complete protocol, combined with PUF, Pedersen commitment based on elliptic curve and zero knowledge proof to complete the design of registration, login, mutual authentication and key negotiation stages.

Registration Phase. Endpoint UD generates the unique identity of the user to be registered and the device, and initiates the registration request to IDP; The IDP returns a random challenge after receiving the registration request. The Endpoint UD terminal then sends the PUF's response and the user's face data to the IDP through the secure communication channel. The IDP terminal generates the classification model and BID based on the proposed method, and uses Pedersen commitment to hide the BID, thus generating the identity token IDT with the digital signature of the IDP terminal used by the user login. The Endpoint UD receives the digitally signed IDT and the encrypted training model output from the IDP terminal and stores it in its own database DB1. Figure 3 illustrates the steps for registration.

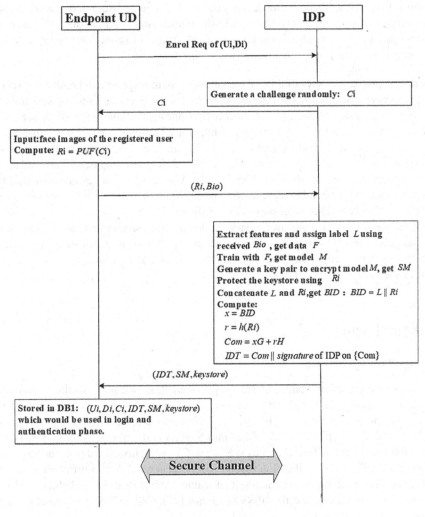

Fig. 3. Enrollment phase

Login and Mutual Authentication Phase. As shown in Fig. 4, this is the user's remote login and authentication phase. At this phase, the Endpoint UD logs in via the IDT with its signature obtained from the IDP, and the input real-time face image. In our scheme, the exchange information involved is not sensitive, does not directly expose the user's biological information, and we do not directly store the user's biological information. After successful login, the database will be automatically updated, and the challenge used will be deleted from database DB1, and then added to database DB2 as data that cannot be reused. With the help of a face for the characteristics of image acquisition equipment can collect, register for the same set of Endpoint UD generate multiple CRPs (Challenge-Response Pairs), the next time you login using the new challenge. This method of automatically updating data not only reduces the pressure of data storage, but also prevents data leakage caused by man-in-the-middle attacks during the authentication process, and relies on the transport layer security protocol to complete the information exchange during the login and authentication phases.

Key Agreement Stage. By combining the key agreement stage and authentication stage, our protocol reaches a higher security level by establishing a secure session key. If there is a malicious party trying to steal information during the communication, sensitive information can be effectively protected. Endpoint UD uses the zero-knowledge proof protocol to prove to the Server that he is a legitimate user. The server then establishes a session key based on the information received during the login and authentication phases, encrypts it and sends it to Endpoint UD. After the session key is established by Endpoint UD, the session key is verified to be the same as the session key of the server, and the session key SK is successfully established.

Proposed scheme realizes user identity authentication which uses the pedersen commitment and zero-knowledge proof, and combines PUF to protect user privacy, successfully resisting common attack methods, such as database attack, replay attack, man-in-the-middle attack, etc.

3   Experiment

3.1   Experiment Settings

To investigate the effectiveness of the proposed method, we perform the experiment based on Matlab 2019B simulation environment to classify and recognize ORL face data on a laptop with Windows 10 operating system. We use a 5-fold cross-validation method to divide each person's 10 faces into 5 equal parts, training four of them each time, and testing the remaining one. When feature extraction is carried out based on eigen faces, different classification results will be generated due to different selection of feature vectors and different amount of training data. In the establishment of the classification model, we use the libSVM library [10], 'V-SVC' SVM type and linear kernel function to classify.

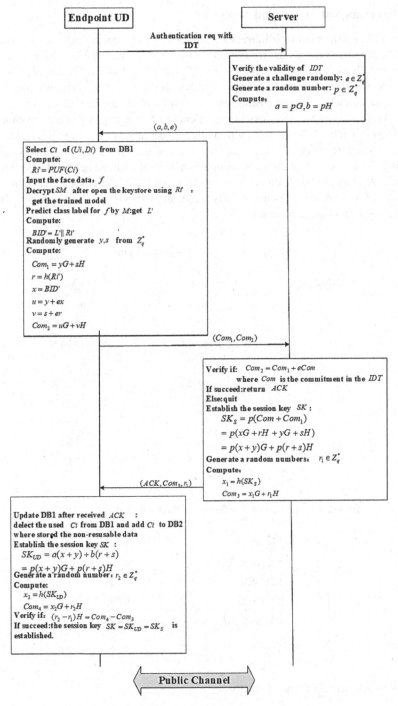

Fig. 4. Login, authentication and key agreement phases

3.2 Experimental Results and Analysis

Table 1 presents the results of different feature vectors for before and after data expansion. the data expansion has a positive impact on the experimental results under the overall situation. When the dimension is between 32 and 44 and above 60, the experimental accuracy is higher after data expansion. Although the accuracy is slightly lower after data expansion when the dimension is between 48 and 56, the overall difference is not large. Therefore, to a certain extent, the method of data expansion can promote the experiment to achieve better recognition effect. According to the experimental results, the first 48 feature vectors of each image are selected to form the feature face. No matter whether the data is expanded, good recognition accuracy can be achieved. As shown in Table 2, we compared the methods and experimental results proposed in this paper with the work of other researchers in the early stage, and it can be seen more intuitively that our method has a higher accuracy.

Table 1. Recognition rate (%) of ORL

Dimension of eigen face space	Before use	After use
28	**96.5**	96.25
32	97	**97.375**
40	97.25	**97.9375**
44	97.75	**97.875**
48	**98.25**	98.1875
52	**98.25**	98.1875
56	**98**	97.875
60	97.75	**97.8125**
64	97.5	**97.875**

Table 2. Accuracy rate of different method using ORL face database

Database used	Method	Accuracy rate (%)
ORL	SVM+PCA [11]	92
ORL	LDA-ED [12]	93.7
ORL	DFA [13]	97.75
ORL	Proposed method	**98.25**

3.3 Revocability and Unlinkability

In order to verify whether the biometrics template protection scheme we proposed meets the revocability and unlinkability, we arbitrarily extract two face images from each user based on the ORL data set to form small data sets (ORL_1 and ORL_2) for performance analysis. As shown in Table 3, we calculated the distributions of genuine, imposter and Mated-imposter scores. It can be observed that the distributions of the Mated-imposter and the genuine are significantly different, while the Mated-imposter distribution closely overlaps the imposter with similar mean values and variances. According to the criteria for evaluating system proposed by Gomez-Barrero et al. [14], this indicates that our proposed scheme supports revocability and unlinkability. When the system is maliciously attacked, the user can re-launch the registration request based on the same biometric to obtain a new BID which has no connection with the original BID. Therefore, our scheme well realizes the security and privacy protection of biometric.

Table 3. Genuine, imposter and mated-imposter distribution for our proposed method

Databases	Genuine		Imposter		Mated-imposter	
	Mean	Variance	Mean	Variance	Mean	Variance
ORL_1	0.512 ± 0.069	0.017 ± 0.014	0.291 ± 0.009	0.023 ± 0.001	0.320 ± 0.013	0.024 ± 0.003
ORL_2	0.591 ± 0.072	0.013 ± 0.006	0.291 ± 0.010	0.022 ± 0.001	0.317 ± 0.010	0.026 ± 0.004

4 Conclusion

In this paper, we propose a novel template protection scheme and two-factor authentication protocol based on multi-service environment using the user's face features and PUF of the device "fingerprint". Specifically, this method not only protects the user biometric template from the perspective of the device, but also can be extended to other biometrics and more classification algorithms. We first use device information to encrypt biological information to generate a cancelable, unique and repeatable biometric identity BID. Then, we combine zero-knowledge proof and Pedersen commitment technology to achieve secure two-way identity authentication and key agreement. And the effectiveness of the proposed scheme is proved by the experiment results.

Acknowledgement. This work was supported in part by the NSFC (Nos. 61976006, 61902003 and 61573023), NSF_AH (Nos. 1808085MF171 and 2108085MF206).

References

1. Kumar, N.: Cancelable biometrics: a comprehensive survey. Artif. Intell. Rev. **53**(11), 3403–34462 (2020)
2. Zheng, Y., Cao, Y., Chang, C.H.: UDhashing: physical unclonable function-based user-device hash for endpoint authentication. IEEE Trans. Industr. Electron. **66**(12), 9559–9570 (2019)
3. Bian, W., Gope, P., Cheng, Y., Li, Q.: Bio-AKA: an efficient fingerprint based two factor user authentication and key agreement scheme. Future Gener. Comput. Syst. **109**, 45–55 (2020)

4. Zhao, J., et al.: A Secure biometrics and PUFs-based authentication scheme with key agreement for multi-server environments. IEEE Access **8**, 45292–45303 (2020)
5. Yang, W., Wang, S., Shahzad, M., Zhou, W.: A cancelable biometric authentication system based on feature-adaptive random projection. J. Inf. Secur. Appl. **58**, 102704 (2021)
6. Arjona, R., Baturone, I.: A dual-factor access control system based on device and user intrinsic identifiers. In: IECON 2016–42nd Annual Conference of the IEEE Industrial Electronics Society, pp. 4731–4736. IEEE (2016)
7. Arjona, R., Prada-Delgado, M.A., Baturone, I., Ross, A.: Securing minutia cylinder codes for fingerprints through physically unclonable functions: an exploratory study. In: 2018 International Conference on Biometrics (ICB), pp. 54–60. IEEE (2018)
8. Liu, Y., Ling, J., Liu, Z., Shen, J., Gao, C.: Finger vein secure biometric template generation based on deep learning. Soft. Comput. **22**(7), 2257–2265 (2017). https://doi.org/10.1007/s00 500-017-2487-9
9. Yang, W., Wang, S., Hu, J., Zheng, G., Yang, J., Valli, C.: Securing deep learning based edge finger vein biometrics with binary decision diagram. IEEE Trans. Industr. Inf. **15**(7), 4244–4253 (2019)
10. Chang, C.C., Lin, C.J.: LIBSVM: a library for support vector machines. ACM Trans. Intell. Syst. Technol. (TIST) **2**(3), 1–27 (2011)
11. Rakshit, P., Basu, R., Paul, S., Bhattacharyya, S., Mistri, J., Nath, I.: Face Detection using Support Vector Mechine with PCA. Available at SSRN 3515989 (2020)
12. Saraswathi, M., Sivakumari, D.S.: Evaluation of PCA and LDA techniques for Face recognition using ORL face database. Int. J. Comput. Sci. Inf. Technol. **1**, 810–813 (2015)
13. Danraka, S.S., Yahaya, S.M., Usman, A.D., Umar, A., Abubakar, A.M.: Discrete firefly algorithm based feature selection scheme for improved face recognition. Comput. Inf. Syst. **23**(2), 23–34 (2019)
14. Gomez-Barrero, M., Galbally, J., Rathgeb, C., Busch, C.: General framework to evaluate unlinkability in biometric template protection systems. IEEE Trans. Inf. Forensics Secur. **13**(6), 1406–1420 (2017)

Realizing Information Flow Control in ABAC Mining

B. S. Radhika[(✉)] [iD] and R. K. Shyamasundar

Indian Institute of Technology Bombay, Mumbai, India
radhikabs@iitb.ac.in

Abstract. Attribute-Based Access Control (ABAC) is an emerging access control model. It is increasingly gaining popularity, mainly because of its flexible and fine-grained access control. As a result, many Role-Based Access Control (RBAC) systems are migrating to ABAC. In such migrations, ABAC mining is used to create ABAC policies from existing RBAC policies. Although ABAC has several advantages, it lacks one of the crucial features required for reliable security, which is information flow control. Due to the complex nature of ABAC policies, it is challenging to analyze the information flows caused by them. In this paper, we address this problem and present an approach for realizing effective information flow control in ABAC systems. With this approach, we can create flow-secure ABAC policies using exiting RBAC policies and associated attributes. With such a flow-secure policy, we can ensure that there are no unintended information flows in the system.

Keywords: Attribute-based access control · ABAC mining · Information flow control

1 Introduction

Access control is one of the key aspects of a secure system. Over the years numerous access models have been proposed. Among them, Discretionary Access Control (DAC), Mandatory Access Control (MAC), and Role-based Access Control (RBAC) are prominent and have been widely used. However, they are proving to be inadequate for the security requirements of the current day systems which require highly dynamic and flexible access control. ABAC addresses this problem by providing authorization based on attributes. Unlike the traditional models, ABAC is not an identity-based model. Instead, it uses attributes of the entities, and the environment and specifies access rules in terms of these attributes. The attributes are nothing but key, value pairs that represent the characteristics or status of the entities or their environment. The access rules can be specified using logical formulas or through enumeration [1]. With the use of attributes, ABAC enables specifying precise access rules.

Although ABAC has several advantages as discussed above, it lacks in one important feature necessary for reliable secure, that is Information Flow Control

W. Meng and M. Conti (Eds.): CSS 2021, LNCS 13172, pp. 107–119, 2022.
https://doi.org/10.1007/978-3-030-94029-4_8

(IFC). IFC is essential to ensure that there are no information leaks in the system and all the accesses are as per the intended security requirements. This would require analysing the information flows allowed by the access policy. However, in the case of ABAC, the policies are generally large and complex. Moreover, they are specified in terms of the attribute values and not the identities. There can be several attributes with large number of possible values. This makes analysing an ABAC policy highly complex. In this paper, we address this problem in an elegant way. Instead of analysing the policy after its creation, we propose a method to ensure flow-security at the policy creation stage and eliminate the need for complex policy analysis.

Specifically, we consider the policy creation in RBAC to ABAC migration where an RBAC policy and associated attribute values are used to generate an equivalent ABAC policy. In the proposed solution, we first analyse the given RBAC policy and make it flow-secure. We then use this flow-secure policy as input for an ABAC mining procedure that generates an equivalent flow-secure ABAC policy.

The rest of the paper is organized as follows: Sect. 2 describes the need for IFC in ABAC. Section 3 discusses the challenges in achieving IFC in ABAC and Sect. 4 presents our approach which can effectively help in creating flow-secure ABAC policies. Section 5 reviews some of the major related works. Finally, Sect. 6 concludes the paper.

2   Motivation

ABAC has several features that make it suitable for most contemporary systems. However, it doesn't focus on IFC which is one of the important security requirements considered essential for reliable security. We demonstrate this with a simple example. Consider a system with two users $u1, u2$ and two objects $o1, o2$. Users have an user attribute $uatt1$ and objects have two attributes $oatt1, oatt2$. Attribute values of the users and objects are listed in Tables 1 and 2 respectively. ABAC policy defined in terms of these attributes is shown in Fig. 1.

<table>
<tr><td colspan="2">Table 1. User attributes</td></tr>
</table>

User	uatt1
u1	uv1
u2	uv2

<table>
<tr><td colspan="3">Table 2. Object attributes</td></tr>
</table>

Object	oatt1	oatt2
o1	ov1	ov2
o2	ov3	ov4

With the permissions available through the given policy, $u1$ can read from $o1$ and write to $o2$ and $u2$ can read from $o2$. The information flows caused by these actions are shown in Fig. 2. From the figure, we can observe that when these accesses are performed, $u2$ can indirectly read the contents of $o1$. This access is not allowed by the policy and hence it is an unauthorized access. Such accesses cause information leaks and pose threat to the system. Therefore they need to

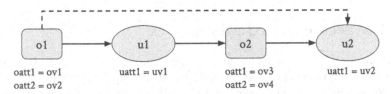

Example Policy

1. $Rule_{read} = \langle (uatt1(u) = uv1 \wedge oatt1(o) = ov1 \wedge oatt2(o)) = ov2) \vee (uatt1(u) = uv2 \wedge oatt1(o) = ov3 \wedge oatt2(o) = ov4) \rangle$
2. $Rule_{write} = \langle (uatt1(u) = uv1 \wedge oatt1(o) = ov3 \wedge oatt2(o)) = ov4) \rangle$

Fig. 1. ABAC policy

Fig. 2. Indirect information flow

be identified and analysed and if they indeed have potential to cause security threat, they need to be eliminated.

In the next section we discuss the challenges involved in information flow analysis of ABAC policies.

3   Information Flow Analysis in ABAC

With attributes-based rules, ABAC policies support flexible and fine-grained access control. However, these benefits come at the cost of complexity and lack of support for auditing. One of the common audit requirement is to know what permissions each user has before the requests are made. This is known as "before the fact audit" [6]. It helps in risk analysis and when an attack manages to compromise a user, auditing/reviewing helps in understanding the scope of the attack and the extent of its damage. It is also necessary to demonstrate compliance to specific regulations or directives. ABAC policies are not well-suited for conducting these audits efficiently. This is mainly because, in ABAC, access rules are not based on the identities of the entities. As a result, determining the set of permissions associated with a user would require enumeration of all the access rules, leading to a significant data retrieval and computation effort. This limitation also affects the information flow analysis capability. Unlike in RBAC, where we can analyse a given policy for information flows and identify indirect flows, policy analysis in ABAC is challenging. Because, such analysis would involve, considering all the attributes (which are usually in large number), each with large number of possible values.

One possible way to achieve IFC in ABAC is to incorporate the flow-security at policy creation time. The process of policy construction is called as *policy engineering*. In the case of ABAC, policy engineering methods can be grouped into two classes, *Top-down* and *Bottom-up*. In the top-down approach, the organization's processes are broken down into tasks and permissions required for each of the tasks are used to derive policy rules. In the Bottom-up approach, also known as *policy mining*, the existing accesses are used to create policy rules.

The existing accesses may be obtained through logs or through existing policies of other models. Policy engineering is the costliest aspect in ABAC implementation. ABAC mining tries to reduce this cost by partially automating the process. As a result, it is usually the preferred method for policy creation. In the next section, we discuss the limitations of ABAC mining with respect to information flow control and how we can address this problem.

4   Flow-Secure ABAC Mining

As mentioned, ABAC policy creation is a costly procedure. ABAC policy mining is used to reduce the cost of policy development by using existing policy to automate the creation of an equivalent ABAC policy. Owing to the fine-granularity and flexibility of ABAC, many existing RBAC systems are migrating to ABAC [16]. In such migrations, policy mining involves using the existing RBAC policy along with the associated attributes to generate an equivalent ABAC policy. The RBAC policy and ABAC mining are formally defined as follows [21]:

Definition 1 (RBAC Policy). *An RBAC policy is a tuple $\langle U, O, R, Op, PERM, UA, PA, RH \rangle$ where U is a set of users, O is a set of objects (resources), R is a set of roles, Op is a set of operations, $PERM \subseteq O \times Op$ is a set of permissions, $UA \subseteq U \times R$ is a user-role assignment relation, $PA \subseteq PERM \times R$ is a permission-role assignment relation, and $RH \subseteq R \times R$ represents role hierarchy.*

Definition 2 (ABAC Policy Mining). *The ABAC policy mining problem is defined as, given an RBAC policy $\pi_0 = \langle U, O, R, Op, PERM, UA, PA, RH \rangle$, user attributes UAT, object attributes OAT, user attribute data du, object attribute data do, find a set of rules Rules such that the ABAC policy $\pi = \langle U, O, Op, UAT, OAT, du, do, Rules \rangle$ is consistent with π_0.*

In this paper, we present an approach to generate flow-secure ABAC policy in such RBAC to ABAC migrations. We explain the method in detail with the help of a running example motivated by the example in [2]. Consider an RBAC configuration as shown in Table 3. It lists the roles, Permission Assignment (PA), and User Assignment (UA). The role hierarchy is $RH = \{(r1, r3)\}$ i.e. $r1$ inherits all the permissions of $r3$. Table 4 and Table 5 give the values of the user attribute (UAT) $uat1$ and object attribute (OAT) $oat1$ respectively. Here, the range of both of these attributes is $\{F, G\}$.

Table 3. RBAC example		
Role	PA	UA
r1	(o1, read) (o3, write)	u1
r2	(o2, write)	u3
r3	(o3, read)	u4, u5
r4	(o1, read) (o3, read) (o3, write)	u2

Table 3. RBAC example

User	uat1
u1	F
u2	F
u3	F
u4	G
u5	G

Table 4. UAT value

Object	oat1
o1	F
o2	F
o3	G

Table 5. OAT value

Note that the RBAC configuration in Table 3 has a transitive flows from $o1$ to $r3$ (when $r1$ or $r4$ reads $o1$, writes it to $o3$ and then $r3$ reads $o3$). If this policy is used as-is for generating ABAC policy, then the resulting policy will also have the transitive flow, making the policy vulnerable to information leaks. To achieve flow-security in ABAC, we need to use flow-secure RBAC policies in ABAC mining. In the proposed method, given an RBAC policy and attribute values, we first identify all the information flows of the RBAC policy and then using this, the RBAC policy is converted into flow-secure policy. Finally, this flow-secure policy is used in ABAC mining to create an ABAC policy that is equivalent to the RBAC policy, thus creating a flow-secure ABAC policy. We describe this procedure in detail as follows:

Step 1: Identifying All the Information Flows: This is done by using the analysis method of [14] that uses an information flow model called Readers-Writers Flow Model (RWFM) [8]. In this model. The entities are assigned RWFM labels and then RWFM checks are applied on each access rule. The two main steps of the method are explained below:

1. Labeling Objects and Roles: The RWFM label of an entity is of the form (R, W), where R corresponds to *readers* and represents the set of roles that can read the entity and W corresponds to *writers* and represents the set of roles that can write or influence the entity.

   In the case of objects, finding the sets of roles that read and write an object is straightforward. The readers set of an object o is obtained from the set of RBAC read rules that include o and writers of o can be obtained from the write rules corresponding to o. Table 6 gives the labels of all the objects in the above example (computed using PA and RH).

Table 6. Labels of objects

Object	Label
o1	$(\{r1, r4\}, \{\})$
o2	$(\{\}, \{r2\})$
o3	$(\{r1, r3, r4\}, \{r1, r4\})$

Table 7. Labels of roles

Role	Label
r1	$(\{r1, r4\}, \{r1, r4\})$
r2	$(\{r1, r2, r3, r4\}, \{r2\})$
r3	$(\{r1, r3, r4\}, \{r1, r2, r3, r4\})$
r4	$(\{r1, r4\}, \{r1, r4\})$

To label the roles, we need to use the labels of the objects. By using the labels of the objects, the readers and writers of a role r are computed such that its readers set is a subset of readers of all the objects it can read and its writers set is a subset of writers of all the objects it can write. To do this, we first initialize the readers and writers sets of a role with the universal set of roles R. For each object o, which is readable by r, we update the $R(r)$ as $R(r) \cap R(o)$. Similarly, when a role r is in the writer set of an object o, we update $W(r)$ as $W(r) \cap W(o)$. The procedure for labelling roles is given in

Algorithm 1. Table 7 gives the labels of all the roles in the example. Label derivation for role $r1$ is given below.

$$R(r1) = \{r1, r2, r3, r4\} \qquad\qquad\qquad\qquad \text{(Initialization)}$$
$$R(r1) = R(r1) \cap R(o1) \cap R(o3) \qquad\qquad (r1 \in R(o1) \wedge r1 \in R(o3))$$
$$R(r1) = \{r1, r2, r3, r4\} \cap \{r1, r4\} \cap \{r1, r3, r4\}$$
$$R(r1) = \{\mathbf{r1, r4}\}$$

$$W(r1) = \{r1, r2, r3, r4\} \qquad\qquad\qquad\qquad \text{(Initialization)}$$
$$W(r1) = W(r1) \cap W(o1) \qquad\qquad\qquad\qquad (r1 \in W(o1))$$
$$W(r1) = \{r1, r2, r3, r4\} \cap \{r1, r4\}$$
$$W(r1) = \{\mathbf{r1, r4}\}$$

Algorithm 1: Label Roles

Input   : RBAC policy (RuleSet) and object labels (L_o)
Output: Labels (L_r) of all the roles in the RuleSet
1 O = getObjects(RuleSet)   ▷ Retrieves all the objects
2 R = getRoles(RuleSet)   ▷ Retrieves all the roles
3 **foreach** $r \in R$ **do**
4 | R(r) = W(r) = R

5 **foreach** $o \in O$ **do**
6 | **foreach** $r \in R(o)$ **do**
7 | | R(r) = R(r) $\cap$ R(o)
8 | **foreach** $r \in W(o)$ **do**
9 | | W(r) = W(r) $\cap$ W(o)

---

2. Applying RWFM Checks: Once the labels of objects and roles are computed, the following checks are applied on each of the RBAC policy rules of the form **r o op** (where $r \in R, o \in O$, and $op \in Op$).
   (a) If $op = read$ check if $W(o) \subseteq W(r)$.
   (b) If $op = write$ check if $R(r) \supseteq R(o)$.
   Any RBAC rule that doesn't satisfy these checks indicates the presence of indirect (transitive) flows i.e. if the condition $W(o) \subseteq W(r)$ fails (in check (a)), then all the roles in $W(o) - W(r)$ can indirectly write to all the objects that r can write. Similarly, if $R(r) \supseteq R(o)$ (in check (b)) fails, then all the roles in $R(o) - R(r)$ can read everything that r can read. We construct RBAC rules corresponding to these indirect accesses and the rule which failed to satisfy the above checks becomes the *cause* for these indirect accesses. The procedure to apply RWFM checks is given in Algorithm 2. On applying this algorithm on the example policy, the checks fail at $(r1, (o3, write))$ and $(r4, (o3, write))$. Both of them cause the indirect flow from $o1$ to $r3$.
   Once all such indirect flows are identified, the rules corresponding to the indirect accesses are added into the rule set and the process (labelling and applying the RWFM checks) is repeated until no new rules are generated. This generates the transitive closure that includes all the multi-level indirections.

The procedure is given in Algorithm 3. In the case of our example, there are no multi-level indirection. Therefore, the analysis stops after one iteration.

Algorithm 2: Check Access Rules

Input : RBAC policy (RuleSet) and labels of objects (L_o) and labels of roles (L_r)

Output: Set of rules corresponding to indirect flow (IndirectRuleSet)

1 IndirectRuleSet = {}
2 **foreach** *rule "r o op"* **do**
3   **if** $op = read$ AND $W(o) \not\subseteq W(r)$ **then**
4   **foreach** $r1 \in (W(o) - W(r))$ **do**
5   **foreach** *o1 which has* $r \in W(o1)$ **do**
6   IndirectRuleSet = IndirectRuleSet $\cup$ {r1 o1 write }

7   **else if** $op = write$ AND $R(r) \not\supseteq R(o)$ **then**
8   **foreach** $r1 \in (R(o) - R(r))$ **do**
9   **foreach** *o1 which has* $r \in R(o1)$ **do**
10   IndirectRuleSet = IndirectRuleSet $\cup$ {r1 o1 read }

Algorithm 3: RBAC Policy Analysis

Input : RBAC policy rules *RuleSet*

Output: Set of all possible indirect allows

1 closure = False
2 **while** *not closure* **do**
3   L_o = labelObjects(RuleSet)   ▷ Computes labels of objects
4   L_r = labelRoles(RuleSet, L_o)
5   IndirectRuleSet = CheckAccessRules(RuleSet, L_o, L_r)
6   **if** *IndirectRuleSet is* $\emptyset$ **then**
7   closure = True
8   **else**
9   RuleSet = RuleSet $\cup$ IndirectRuleSet

Step 2: Creating a Flow-secure Policy: From the previous step, we can not only identify all the indirect flows, but can also identify the rules causing those indirections. With these details, we can create a flow-secure RBAC policy by applying one of the following measures for each of the indirections:

– Ignore: Generally, not all the indirect flows are a security concern. Certain indirect flows can be ignored and can be allowed to be part of the policy, if the policy writer decides that the flow is not a security threat. One possible reason for such decision could be due to the fact that some roles are trusted and are believed not likely to cause indirections.

- Remove: A straightforward measure is to remove the indirections. This is done by deleting at least one rule from the path that leads to a particular indirection. Note that deleting rules from the policy may affect the functionality of roles. This has to be considered while removing the rules.
- Add: In the case of some indirections, the corresponding rules can be explicitly added to the policy. This can be done if the policy writers considers the indirect access to be safe and necessary for functioning of certain roles.

Using the above measures, we can convert a given policy into a flow-secure version where there are no unintended flows. In the case of our example policy, suppose the the role $r1$ is more trusted and can be relied on not to write the contents of $o1$ to $o3$. Then we can ignore the indirect flow through $r1$. Whereas the indirection caused through the regular role $r4$ needs to be removed. After removal of $(r4, (o3, write))$, permissions of users will be as shown in Table 8. With this, no access in the policy causes an unintended flow.

Table 8. User permissions

User	Permissions
$u1$	$(o1, read)$ $(o3, read)$ $(o3, write)$
$u2$	$(o1, read)$ $(o3, read)$
$u3$	$(o2, write)$
$u4$	$(o3, read)$
$u5$	$(o3, read)$

Table 9. Partitions

P1 uat1=F, oat1=F
$(u1, o1)$
$(u1, o2)$
$(u2, o1)$
$(u2, o2)$
$(u3, o1)$
$(u3, o2)$

P2 uat1=F, oat1=G
$(u1, o3)$
$(u2, o3)$
$(u3, o3)$

P3 uat1=G, oat1=F
$(u4, o1)$
$(u4, o2)$
$(u5, o1)$
$(u5, o2)$

P4 uat1=G, oat1=G
$(u4, o3)$
$(u5, o3)$

Step 3: Generating Equivalent ABAC Policy: Once we have a flow-secure RBAC policy, we can use it to generate an equivalent ABAC policy. We do this by creating conflict-free partitions [2]. The detailed procedure is explained below:

- **Create Partitions:** The first step is to divide the set $U \times O$ into a partition set $P = \{P_1, P_2, ..., P_n\}$ such that all the attribute values of every tuple in a partition are same. In the case of our example, four such partitions are created. The partitions and the corresponding attribute values are shown in Table 9.

ABAC Policy 1
1. $Rule_{read} = \langle(uat1(u) = G \wedge oat1(o) = G)\rangle$
2. $Rule_{write} = \langle\rangle$

Fig. 3. ABAC policy for partitions P3 and P4

- **Make Partitions Conflict-free:** A partition is said to be conflict-free if every tuple in the partition has same set of permissions as per the RBAC policy. Such partitions can be uniquely identified by their attributes and this can be used to create ABAC rules. Notice that among the partitions in Table 9, $P3$ and $P4$ are conflict-free, because all their tuples have same permissions. As a result, the partitions' attributes are sufficient to uniquely identify them and can specify the ABAC rules as shown in Fig. 3. On the other hand, the partitions $P1$ and $P2$ are not conflict-free. In the case of $P1$, users $u1$ and $u2$ can read $o1$ but $u3$ cannot. Such partitions need to be further split into smaller partitions where all the tuples have same authorizations. i.e. a partition P_i is split into $\{S_1, S_2, ..., S_m\}$ such that each resulting partition is conflict-free. This is done by first splitting the set of users U_i and set of objects O_i of partition P_i. U_i is split into $\{U_{i1}, U_{i2}, ..., U_{ik}\}$ such that all the users in any U_{ij} has same authorization with respect to all the objects. Similarly, O_i is split into $\{O_{i1}, O_{i2}..., O_{il}\}$ such that all the objects in any O_{ij} can be accessed in the same way by all the users. Then conflict-free partitions are created as $\{U_{i1}, U_{i2}..., U_{ik}\} \times \{O_{i1}, O_{i2}..., O_{il}\}$. Note that such partitions cannot be uniquely identified by their input attribute values alone. Therefore, we add new attributes so that these partitions can be uniquely identified and can be used to specify access rules.
  From Table 8, we can see that all three users of partitions $P1$ and $P2$ have different set of permissions and the two objects in $P1$ have different set of readers and writers. As a result, users and objects in both partitions $P1$ and $P2$ are split into $\{u1\}\{u2\}\{u3\}$ and the objects set of $P1$ is split into $\{o1\}\{o2\}$. Consequently, each element in these partitions form a separate conflict-free partition.
- **Assigning New Attributes:** In this step, we add a new user attribute *urole* and a new object attribute *orwfm*. Value of *urole* attribute of a user specifies the set of roles that can be assigned to the user as per the RBAC configuration. Value of *orwfm* attribute of an object corresponds to its RWFM label. The attribute values of the users is shown in Table 10 and the new object attribute values are as per Table 6 (except for $o3$, whose writers set is updated to $\{r1\}$).

Table 10. User role attribute

User	urole
u1	$\{r1, r3\}$
u2	$\{r4\}$
u3	$\{r2\}$
u4	$\{r3\}$
u5	$\{r3\}$

– **Generate Rules:** All the conflict-free partitions created in the previous steps can be uniquely identified by the combination of the input attributes and the newly added attributes. Using them we can create ABAC rules. After creating such rules for partitions $P1$ and $P2$ and adding them to Policy 1, we get the complete policy as shown in Fig. 4. The rules generated this way do not contain any unintended flows, thus making the system flow-secure.

ABAC Policy 2

1. $Rule_{read} = \langle\langle(uat1(u) = G \wedge oat1(o) = G)\rangle\rangle \vee$
   $(uat1(u) = F \wedge oat1(o) = F \wedge urole(u) \in \{\{r1, r3\}, \{r4\}\} \wedge orwfm(o) = (\{r1, r4\}, \{\})) \vee$
   $(uat1(u) = F \wedge oat1(o) = G \wedge urole(u) \in \{\{r1, r3\}, \{r4\} \wedge orwfm(o) = (\{r1, r3, r4\}, \{r1\})))$
2. $Rule_{write} = \langle\langle(uat1(u) = F \wedge oat1(o) = F \wedge urole(u) = \{r2\} \wedge orwfm(o) = (\{\}, \{r2\})) \vee$
   $(uat1(u) = F \wedge oat1(o) = G \wedge urole(u) = \{r1, r3\} \wedge orwfm(o) = (\{r1, r3, r4\}, \{r1\})))$

Fig. 4. ABAC policy

With the above method, we eliminate the need for flow analysis of an ABAC policy to ensure its flow security. The method uses the simple and elegant approach of using a flow-secure policy as input for ABAC mining. Though we have used mining based on conflict-free partitions, it is possible to use any other mining solution as long as the resultant ABAC policy is equivalent to the RBAC policy.

5   Related Works

Over the years, several ABAC models have been proposed [19,22], each aiming to provide flexible and fine-grained access control. However, since ABAC is relatively new model, there has been less work on information flow control. Jin et al. [7] presented a unified ABAC model called $ABAC_\alpha$. It provides a formal model that can cover the traditional models – DAC, MAC, and RBAC. Here the model can use MAC labels as attributes and apply the MAC rules based on these attributes. However, the labels have to be assigned statically which makes the model restrictive. Our method provides more flexibility by using the RWFM model.

In the case of RBAC, relatively more work has been done regarding IFC. Nyanchama and Osborn [10] provided general rules for RBAC to MAC simulation. Sandhu [15] and Osborn et al. [13] presented a method to emulate lattice-based access control using two role hierarchies. Nyanchama and Osborn [11] presented a role-graph model that can address the conflict of interest with respect to information flows. This model is useful in understanding the information flows in RBAC systems. Osborn [12] and Gofman et al. [5], also focus on information flow control in RBAC. They generate information flow graphs using different techniques. The graphs show information flows at different granularities and aim to answer flow-related queries. Tuval and Gudes [17] attempt to analyze the information flows in a given RBAC configuration and try to resolve information flow conflicts by creating canonical groups. These flow-graph-based solutions work for only smaller systems with a few roles and objects. As the number of roles and objects increase, the analysis becomes complex and cumbersome. The method used in this paper is scalable and more effective in analysing the indirect flows and extracting related details.

ABAC Mining is recently being explored widely by researchers. Vaidya et al. [18] and Das et al. [3] use migration based approach which enables an organization to use an existing policy from another organization having similar attributes and their values. The method in [3] enables an organization using a traditional access control model to adapt to the existing ABAC policy of a similar organization by an optimal assignment of subject attribute values in the presence of attribute hierarchy and environmental conditions. Gautam et al. [4] use a constrained mining approach for policy engineering. Xu and Stoller [20] use a log based approach whereas Mocanu et al. [9] use both log and deep learning. The method in our approach uses an existing RBAC policy and associated attributes as input. The method is simple and effective.

6   Conclusions

ABAC is an emerging access control model. Although it has several advantages such as fine-granularity and flexibility, it lacks support for information flow analysis. In this paper we have proposed an approach that can create flow-secure ABAC policies by using existing RBAC policies. This eliminates the need for flow analysis of ABAC policy. Thus our method provides an elegant solution to realize information flow control in ABAC systems thereby making them more reliable.

References

1. Biswas, P., Sandhu, R., Krishnan, R.: Label-based access control: an ABAC model with enumerated authorization policy. In: Proceedings of ACM International Workshop on Attribute Based Access Control, pp. 1–12 (2016)
2. Chakraborty, S., Sandhu, R., Krishnan, R.: On the feasibility of RBAC to ABAC policy mining: a formal analysis. In: Proceedings of International Conference on Secure Knowledge Management In Artificial Intelligence Era, pp. 147–163 (2019)

3. Das, S., Sural, S., Vaidya, J., Atluri, V.: Policy adaptation in attribute-based access control for inter-organizational collaboration. In: Proceedings of IEEE International Conference on Collaboration and Internet Computing, pp. 136–145 (2017)
4. Gautam, M., Jha, S., Sural, S., Vaidya, J., Atluri, V.: Poster: constrained policy mining in attribute based access control. In: Proceedings of ACM Symposium on Access Control Models and Technologies, (SACMAT) (2017)
5. Gofman, M.I., Luo, R., Solomon, A.C., Zhang, Y., Yang, P., Stoller, S.D.: RBAC-PAT: a policy analysis tool for role based access control. In: Proceedings of International Conference on Tools and Algorithms for the Construction and Analysis of Systems (TACAS), pp. 46–49 (2009)
6. Hu, V.C., et al.: Guide to attribute based access control (abac) definition and considerations (draft). NIST Spec. Publ. 800(162), 1–54 (2013)
7. Jin, X., Krishnan, R., Sandhu, R.S.: A unified attribute-based access control model covering dac, MAC and RBAC. In: Proceedings of IFIP WG 11.3 Working Conference on Data and Applications Security (DBSec), pp. 41–55 (2012)
8. Kumar, N.V.N., Shyamasundar, R.K.: A complete generative label model for lattice-based access control models. In: Proceedings of International Conference on Software Engineering and Formal Methods (SEFM), pp. 35–53 (2017)
9. Mocanu, D., Turkmen, F., Liotta, A., et al.: Towards ABAC policy mining from logs with deep learning. In: Proceedings of the 18th International Multiconference, ser. Intelligent Systems (2015)
10. Nyanchama, M., Osborn, S.L.: Modeling mandatory access control in role-based security systems. In: Proceedings of IFIP WG 11.3 Working Conference on Data and Applications Security (DBSec), pp. 129–144 (1995)
11. Nyanchama, M., Osborn, S.L.: The role graph model and conflict of interest. ACM Trans. Inf. Syst. Secur. (TISSEC) 2(1), 3–33 (1999)
12. Osborn, S.L.: Information flow analysis of an RBAC system. In: Proceedings of ACM Symposium on Access Control Models and Technologies (SACMAT), pp. 163–168 (2002)
13. Osborn, S.L., Sandhu, R.S., Munawer, Q.: Configuring role-based access control to enforce mandatory and discretionary access control policies. ACM Trans. Inf. Syst. Secur. 3(2), 85–106 (2000)
14. Radhika, B.S., Kumar, N.V.N., Shyamasundar, R.K.: Towards unifying RBAC with information flow control. In: Proceedings of ACM Symposium on Access Control Models and Technologies (SACMAT), pp. 45–54 (2021)
15. Sandhu, R.S.: Role hierarchies and constraints for lattice-based access controls. In: Proceedings of European Symposium on Research in Computer Security (ESORICS), pp. 65–79 (1996)
16. Sandhu, R.S.: The authorization leap from rights to attributes: maturation or chaos? In: 17th ACM Symposium on Access Control Models and Technologies (SACMAT), pp. 69–70 (2012)
17. Tuval, N., Gudes, E.: Resolving information flow conflicts in RBAC systems. In: Proceedings of IFIP WG 11.3 Working Conference on Data and Applications Security (DBSec), pp. 148–162 (2006)
18. Vaidya, J., Shafiq, B., Atluri, V., Lorenzi, D.: A framework for policy similarity evaluation and migration based on change detection. In: Proceedings of International Conference on Network and System Security (NSS), pp. 191–205 (2015)
19. Wang, L., Wijesekera, D., Jajodia, S.: A logic-based framework for attribute based access control. In: Proceedings of the ACM Workshop on Formal Methods in Security Engineering (FMSE), pp. 45–55 (2004)

20. Xu, Z., Stoller, S.D.: Mining attribute-based access control policies from logs. In: Proceedings of IFIP WG 11.3 Working Conference on Data and Applications Security (DBSec), pp. 276–291 (2014)
21. Xu, Z., Stoller, S.D.: Mining attribute-based access control policies. IEEE Trans. Dependable Secur. Comput. **12**(5), 533–545 (2015)
22. Zhang, X., Li, Y., Nalla, D.: An attribute-based access matrix model. In: Proceedings of the ACM Symposium on Applied Computing (SAC), pp. 359–363 (2005)

Weak Password Scanning System
for Penetration Testing

Bailin Xie[1,2(✉)], Qi Li[1,2], and Hao Qian[1,2]

[1] School of Information Science and Technology, Guangdong University of Foreign Studies,
Guangzhou, China
bailinxie@gdufs.edu.cn
[2] School of Cyber Security, Guangdong University of Foreign Studies, Guangzhou, China

Abstract. Nowadays, many network security related personnel are accustomed to using simple passwords or default passwords set by system. Based on this kind of weak password vulnerabilities, the hackers can gain access to the systems easily. Weak password scanning is an important part of penetration testing. In order to enable penetration testers to discover weak passwords in the system more conveniently, this paper proposes a system for weak password scanning. This system includes five modules, namely the interface module, data reading processing module, IP address survival detection module, task scheduling module, and the weak password scanning plugin module. Furthermore, this system is developed based on the Go language, which has the characteristics of supporting high concurrency from the language level. We test this system by using the environment built by Docker. The experimental results validate the effectiveness of this system. In the actual penetration testing, this system can save a lot of time and energy for personnel, and has a certain practical value.

Keywords: Weak password · Penetration testing · Scanning system · Go language

1 Introduction

With the rapid development and popularization of the Internet, people gradually rely on the Internet. Especially during the COVID-19 pandemic [17], people have adopted lifestyles such as "cloud to work", "cloud learning", and "cloud shopping". This pandemic makes social operation and business contacts more dependent on the Internet. According to the 45th "Statistical Report on China's Internet Development Status", as of March 2020, the number of Internet users in China has reached 903 million, and the Internet penetration rate has reached 64.5%.

With the rapid growth of Internet users, more and more websites have been put into use. According to the survey report of NetCraft, the number of global websites has reached 126 million by February 2020. Currently, the websites in all over the world are facing severe network security threats. For example, in May 2017, WannaCry ransomware swept the world. The servers of some universities and medical institutions were

W. Meng and M. Conti (Eds.): CSS 2021, LNCS 13172, pp. 120–130, 2022.
https://doi.org/10.1007/978-3-030-94029-4_9

infected, causing their websites to be paralyzed and unable to be accessed normally. Ransomware [1] usually uses weak password vulnerabilities in the system to spread. For example, the ransomware Lucky uses a combination of weak password vulnerabilities and Window SMB vulnerabilities [2] for rapid attack and spread.

Weak password refers to the password that is easily guessed by others or cracked by cracking tools. There are mainly three types of weak passwords: (1) short passwords. The passwords with a length of less than 6 digits are usually considered as short passwords; (2) the password set by default. Some developers will set an initial password for the administrator when developing the system, and recommend to modify the initial password of the administrator user after the first login; (3) set a relatively simple password, such as the keyboard password "qwertyuiop", "1q2w3e4r" and other passwords [3, 4].

In recent years, as the value of digital currency [5, 6] continues to rise, the activities of mining Trojan horse are more and more frequent. Intranet is an ideal environment for mining Trojan horse. In the intranet, there are thousands of hosts, which is the hardest hit by weak password vulnerabilities. Weak password vulnerability has the characteristics of high threat and easy to be exploited. Some administrators do not change the default passwords of the system and service settings, or use the passwords that have been publicly leaked on the network as the system passwords. These passwords are useless and give hackers the opportunity to enter the intranet. Once a host in the intranet is captured by the mining Trojan horse, it will use the built-in weak password dictionary to brute force the passwords of other hosts in batches, and then combine with high-risk vulnerabilities, move horizontally and continuously try to infect more hosts in the network. Finally, it will cause serious damage to the internal network environment and bring huge losses to the enterprise.

In order to better protect the network and system security, penetration testing came into being. Penetration testing is to evaluate the system and network security from the perspective of the attacker by simulating the attack methods of hackers, find out the loopholes in the system and propose a repair scheme. Weak password scanning is an important part of penetration testing. In order to make penetration testers find weak password vulnerabilities in networks and systems faster and more conveniently, this paper proposes a weak password scanning system.

The remainder of the paper is organized as follows. Section 2 reviews recent studies on penetration testing and weak password scanning. In Sect. 3, we introduce the principle of weak password scanning. Section 4 presents the design and implementation of the system. Evaluation and experimental results are given in Sect. 5. Finally, we conclude the paper in Sect. 6.

2 Related Work

According to the tester's familiarity with the target system and network, penetration testing can be divided into black-box testing, white-box testing and gray-box testing. Black-box testing is a testing method in which the network and internal structure of the target system are not known to the tester; white-box testing refers to penetration when testers have a certain understanding of the target system version, source code and network topology; gray-box testing is a combination of black-box testing and white-box testing.

At present, much previous research on penetration testing mainly focuses on methods and standards. For example, Xu et al. [7] proposed a penetration testing method for cyber-physical system based on attack graph, Le et al. [8] proposed a penetration testing method by using blind SQL injection based on second-order fragment and reassembly, and Zhou et al. [9] designed a penetration testing tool for industrial control systems based on shell interaction technology. Halfond et al. [10] proposed to improve the performance of penetration testing by combining static and dynamic analysis. Antunes et al. [11] conducted a comparative analysis of several popular penetration testing tools for web systems and pointed out their shortcomings. Al-Ahmad et al. [12] analyzed the penetration testing of mobile cloud computing applications and pointed out possible research in the future. Zhou et al. [13] proposed an attack planning algorithm from the perspective of the attacker, which based on network information gain and utilized this algorithm to achieve automatic penetration testing. Tian et al. [14] proposed an attack model based on penetration testing, and then used this model to detect SQL injection vulnerabilities. Shah et al. [15] conducted a comparative analysis of some existing penetration testing techniques.

However, few researchers conducted research on weak password scanning method in penetration testing. Therefore, this paper focuses on the weak password scanning in penetration testing, and proposes a weak password scanning system.

3   Principle of Weak Password Scanning

The procedure of the weak password scanning system proposed in this paper is shown in Fig. 1. After starting the system, the scanning options can be set according to the instructions of the menu, including the number of scanning coroutines, the file of target IP address, user name dictionary and password dictionary. Then, read the data from the IP address file, user name dictionary file and password dictionary file, which are set in the parameter configuration. After that, perform simple processing on the acquired data, including removing the null value, identifying the service in the IP address, removing the IP address corresponding to the unsupported service, and detecting processed IP addresses to remove unreachable IP addresses. Finally, combining the user names and passwords in the dictionary one by one, and trying to log in the corresponding service in each IP address through a large amount of concurrency, so as to detect the weak password used by the target.

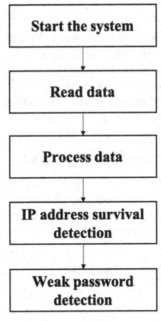

Fig. 1. System flow.

4 Design and Implementation

The weak password scanning system proposed in this paper is developed based on the Go language, which supports concurrency from the language level and is simple to use. The system consists of five parts, namely the interface module, data read and process module, IP address survival detection module, task scheduling module, and the weak password scanning plugin module. The structure of the system is shown in Fig. 2.

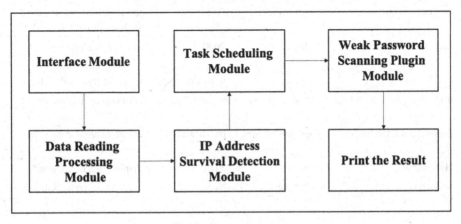

Fig. 2. System structure.

4.1 Interface Module

The main function of this module is to provide an operation interface for the system. The contents displayed on the interface are the system description and function menu, in which system description includes the system name, usage introduction, author information and version information, as shown in Table 1. Function menu includes command options and parameter settings, as well as descriptions of corresponding parameters, as shown in Table 2 and Table 3 respectively.

Table 1. System description.

Name	Content	Description
system name	QPwd	display system name
usage	./QPwd command	startup command
version	2020	update date
author	fakeuser@fakemail.com	author information

Table 2. Command options of function menu.

Function	Command	Description
scan	scan	scanning, parameters need to be set
help	help	print function menu

Table 3. Parameter setting of function menu.

Name	Parameter	Default	Description
timeout value	--timeout, -t	5(s)	threshold for judging the survival of the target IP
number of coroutines	--thread, -n	2000	number of started coroutines
IP address file	--ipfile, -i	ipfile.txt	store the target IP address
user name dictionary file	--username, -u	username.dic	common user name collection
password dictionary file	--password, -p	password.dic	common weak password set
Help menu	--help, -h	None	display function menu
version	--version, -v	None	output version information

4.2 Data Reading Processing Module

The main function of the data reading processing module is to read the username dictionary file, password dictionary file and IP address file, and output the processed formatted data. The contents of the username dictionary file and the password dictionary file are the user names and passwords commonly used in the network system. The content of the IP address file is the result of port scanning. We use the port scanning tool to scan the target system and network, and simply process on each scanning result. Only the IP address, port, and service name are reserved. Furthermore, we save them as an IP address file in the format of "IP:port|service name".

This module mainly uses the os, bufio, strconv and strings libraries in the Go language. When opening a file to read data, it uses the os.Open() function to create a handle named file. If the file does not exist or an error occurs, the formatted error message is output by the logger.Log.Fatalf() function. After reading the file, it uses the defer file.Close() function to close the file. The format supported by each line is: IP:Port|Service.

When processing an IP address file, it use the bufio.NewScanner() and bufio.ScanLines() functions to read each line of data in the file one by one, and use the strings.TrimSpace() function to remove the blanks in each line. This module uses the strings.Split() function to split the string, first split each line into two parts according to to "|", namely IP: port and service. The service part is converted to uppercase using the strings.ToUpper() function and saved as the variable serviceName. Then divide the "IP: port" into two parts according to ":" and save them as variables IP and port. The port variable needs to be converted to an integer type using the strconv.Atoi() function. Finally, judge whether it is a supported service according to the serviceName, and reserve the supported services. For unsupported services, output unsupported prompt information on the terminal.

When processing the user name dictionary file and the password dictionary file, we use the bufio.NewScanner() and bufio.ScanLines() functions to read the username and password of each line one by one, and use the strings.TrimSpace() function to remove the blanks in each line. We judge whether the row is empty, if it is not, it will be stored in the slice. Finally, the processed slices are output.

4.3 IP Address Survival Detection Module

The main function of the IP address survival detection module is to remove unreachable IP. Before scanning the weak password, the IP address and port obtained from the data reading processing module are detected to determine whether the IP address is reachable. Then, the unreachable addresses are removed, and the reachable IP address slices are used as output. This module can reduce the error rate and unnecessary workload in the process of weak password scanning.

This module mainly uses the sync and net libraries in the Go language, use the check() function to receive the IP slice and return the survival state and the survival IP slice. The initial value of the variable alive is false. We use the net.DialTimeout() function to make a TCP connection to each piece of data in the IP slice. This function returns the connection handle and error information. If the connection does not produce an error, record the value of the variable alive as true.

4.4 Task Scheduling Module

Task scheduling module is divided into two parts: creating tasks and processing tasks. The task creation calculates the data which processed by the data reading processing module and the IP address survival detection module to generate task slices, and then passes them to the processing task. The total number of tasks generated = the number of user names × the number of passwords × the number of IP.

The processing task creates a channel, starts the coroutine, and sends the task into the channel. Then, identifies the service according to each task. Finally, sends it to the corresponding weak password scanning plugin.

This module mainly uses the sync and time libraries in the Go language. When creating a task, use the CreateJobs() function to receive the IP slice, user name slice, and password slice output from the IP address survival detection, and return the task slice and the number of tasks. The task is a combination of IP address, port, service, username and password.

4.5 Weak Password Scanning Plugin Module

The main function of the weak password scanning plugin module is to detect the weak password of the supported services. It supports the detection of five common services including FTP, SSH, MySQL, Redis and MongoDB. According to the service name, it scans the weak password with the corresponding plugin. Through the combination of username and password, try to log in to the service corresponding to each IP address and port, thereby detects the weak password used by the target. This module is developed in the form of plugin, which is convenient to add scanning plugins for other services in subsequent use. Thus, this can improve the flexibility and scalability of the system. This module mainly uses third-party operation libraries of FTP, SSH, MySQL, Redis, and MongoDB under the Go language.

FTP Scanning Plugin Implementation. We use the ScanFtp function to receive tasks, return errors and scanning results. This plugin uses the ftp.DialTimeout() function to create a connection handle named conn to connect the IP and port in the task, then uses the conn.Login() function to try to log in, in which the username and password are used as input. If there is no error, the variable result.Weak will be set as true, and this password will be marked as a weak password. Finally, the connect is disconnected by calling the defer conn.Quit() function.

SSH Scanning Plugin Implementation. This plugin firstly declares a variable named "config", and configures the SSH login parameters through &ssh.ClientConfig{}, including username, authentication method, timeout period and server verification. The authentication method can use public key and private key for authentication, here we use Auth: []ssh.AuthMethod for password authentication. Then we use the ssh.Dial() function to create a connection handle named client, and connect through the config variable. If there is no error, the variable result.Weak will be set as true, and this password will be marked as a weak password. And finally we use the defer client.Quit() function to disconnect.

MySQL Scanning Plugin Implementation. This plugin firstly declares the dataSourceName variable and sets the format and parameters for connecting to MySQL. Then we use the sql.Open() function to create a connection handle named "db", and open the default database named "mysql". In addition, we use the db.Ping() function to verify the connection. If there is no error, the variable result.Weak will be set as true, and this password will be marked as a weak password. Finally, we use the defer db.Close() function to disconnect.

Redis Scanning Plugin Implementation. This plugin firstly declares a variable named "opt", and sets the format and parameters for connecting to Redis through the redis.Options{} structure. Then we use the redis.NewClient(&opt) function to establish a connection and return a connection handle named client. If there is no error, the variable result.Weak will be set as true, and this password will be marked as a weak password. Finally, we use the defer client.Close() function to disconnect.

MongoDB Scanning Plugin Implementation. This plugin firstly declares the URL variable and sets the format and parameters for connecting to MongoDB. Then we use the function named mgo.DialWithTimeout() to return a connection handle named session. If there is no error, the variable result.Weak will be set as true, and this password will be marked as a weak password. Finally, the defer session.Close() function will be used to disconnect.

5 System Testing and Result Analysis

We tested the weak password scanning system proposed in this paper by using Docker [16] to build a test environment, in which the version of Docker is Docker Desktop for Mac 2.2.0.3 (42716). Docker is a tool that can realize the virtualization of the operating system layer. By packaging the application as a virtual container isolated from the host, it avoids the problem of unnecessary resource occupation caused by starting the virtual machine and the problems that may be encountered in the process of configuring the environment. In the experiments, we download the images corresponding to the five services from Docker Hub, and build containers by these. When starting the container, the username and password of the five services will be set at first. Through port mapping, the ports opened by the services in the container are mapped to the ports of the host. The test environment information is shown in Table 4.

The operating system used for the experiments is macOS 10.15.3, the version of Golang SDK is 1.13.8, and the terminal uses ZSH, which the version is 5.7.1 ($\times$86_64-apple-darwin19.0).

First, we use the port scanning tool to scan, get the information shown in Table 5, and then import this information into the IP list. Since our test is done locally, the IP address is uniformly set to 127.0.0.1, the port and protocol are set according to the scanning results. In addition, the top 10 most used usernames on the network are set as the username dictionary, and the weak password dictionary is composed by using the top 50 most common weak passwords on the network. The number of coroutines is used the default value 2000. The scanning results of weak passwords are shown in Table 6.

Table 4. Test environment information.

Container	Used images	Service username/password	Open port
FTP	teezily/ftpd	ftp_U53R/87#dn@A8nv	11021
SSH	hermsi/alpine-sshd	root/123321	11022
MySQL	mysql:5.6	root/123456	3306
Redis	redis	admin/redis123	6379
MongoDB	mongo	admin/123456	27017

Table 5. Port scanning results.

Open state	Port type	Port number	Remark
Open	TCP Port	3306	Mysql
Open	TCP Port	6379	Redis
Open	TCP Port	11021	None
Open	TCP Port	11022	OpenSSH
Open	TCP Port	27017	MongoDB

It can be seen from Table 6 that the weak password scanning system proposed in this paper scanned the five test services and successfully found the weak passwords in the service systems of SSH, Redis, MySQL and MongoDB, but failed to find the passwords in the FTP service. From the perspective of the attacker, the password set by the FTP service system is not a traditional weak password, and has not been included in the weak password dictionary used in the test, so it cannot be found. However, in the actual penetration testing, the success rate of the system can be improved by using more weak password dictionaries, or using weak password dictionaries made through social engineering.

Table 6. Weak password scanning results.

Service	Scanning result	User name/Password
FTP	No	Null
SSH	Yes	root/123321
Redis	Yes	admin/redis123
MySQL	Yes	root/123456
MongoDB	Yes	admin/123456

The performance of the system is shown in Table 7. Since the number of usernames and passwords for each service included in the username dictionary and the weak password dictionary are the same, the scanning time for each service is basically the same. The system successfully found out the weak passwords in SSH, Redis, MySQL and MongoDB service, and the accuracy rate was 100%. However, the password in the FTP service could not be found.

Table 7. System performance.

Service	Scanning time (s)	Accuracy
FTP	0.11	0
SSH	0.12	100%
Redis	0.11	100%
MySQL	0.12	100%
MongoDB	0.12	100%

It can be seen from the above experiment that this system can replace complicated manual testing in actual penetration testing, which can save a lot of time and energy for penetration testers, and our experiments demonstrate the practical value of our system.

6   Conclusion

In this paper, we propose a weak password scanning system for penetration testing. This system consists of five parts, namely the interface module, data reading and processing module, IP address survival detection module, task scheduling module, and the weak password scanning plugin module. In addition, this system is developed based on Go language, which supports high concurrency from the language level. We used Docker to build a test environment and done an experiment for this system. The experimental results demonstrate the practical value of our system.

At present, our system needs to rely on the results of external port scanning, which is not smart enough. Therefore, the future step of this work is to expand the port scanning and fingerprint recognition functions to make the system more intelligent.

Acknowledgments. This work is supported by the Guangdong Basic and Applied Basic Research Foundation (Grant No. 2018A0303130045), the Science and Technology Program of Guangzhou (Grant No. 201904010334).

References

1. Yu, H., Peng, G., Cai, K.: Research on file recovery method against ransomware using hybrid pattern cryptographic system. Comput. Eng. Appl. **55**(10), 96–102 (2019)

2. Li, Y., Huang, C., Wang, Z., Yuan, L., Wang, X.: Survey of software vulnerability mining methods based on machine learning. J. Softw. **31**(07), 2040–2061 (2020)
3. Spafford, E.H.: Preventing weak password choices. In: Proceedings of the 14th National Computer Security Conference, pp. 446–455. Springer, Heidelberg (1992)
4. Weber, J.E., Guster, D., Safonov, P., Schmidt, M.B.: Weak password security: an empirical study. Inf. Secur. J. Glob. Perspect. **17**(1), 45–54 (2008)
5. Zhang, Z., Wang, M.: Survey on blockchain wallet scheme. Comput. Eng. Appl. **56**(06), 28–38 (2020)
6. Xie, K.: Study on evolution of digital currency based on blockchain. Appl. Res. Comput. **36**(07), 1935–1939 (2019)
7. Xu, B., He, G.: Penetration testing method for cyber-physical system based on attack graph. Comput. Sci. **45**(11), 143–148 (2018)
8. Le, D., Gong, S., Wu, S., Liu, W.: Penetration test method using blind SQL injection based on second-order fragment and reassembly. J. Commun. **38**(S1), 77–86 (2017)
9. Zhou, W., Yang, W., Wang, X., Ma, B.: Research on penetration testing tool for industrial control system. Comput. Eng. **45**(08), 92–101 (2019)
10. Halfond, W.G.J., Choudhary, S.R., Orso, A.: Improving penetration testing through static and dynamic analysis. Softw. Test. Verification Reliab. **21**(3), 195–214 (2011)
11. Antunes, N., Vieira, M.: Penetration testing for web services. Computer **47**(2), 30–36 (2014)
12. Al-Ahmad, A.S., Kahtan, H., Hujainah, F., Jalab, H.A.: Systematic literature review on penetration testing for mobile cloud computing applications. IEEE Access **7**, 173524–173540 (2019)
13. Zhou, T., Zang, Y., Zhu, J., Wang, Q.: NIG-AP: A new method for automated penetration testing. Front. Inf. Technol. Electron. Eng. **20**(9), 1277–1288 (2019)
14. Tian, W., Yang, J.F., Xu, J., Si, G.N.: Attack model based penetration test for SQL injection vulnerability. In: Proceedings of the 2012 IEEE 36th Annual Computer Software and Applications Conference Workshops, pp. 589–594. IEEE Computer Society, Washington D.C., United States (2012)
15. Shah, S., Mehtre, B.M.: An overview of vulnerability assessment and penetration testing techniques. J. Comput. Virol. Hacking Tech. **11**(1), 27–49 (2014). https://doi.org/10.1007/s11416-014-0231-x
16. Wang, J., Hu, W., Zhang, Y., et al.: Trusted container based on docker. J. Wuhan Univ. (Sci. Edn.) **63**(2), 102–108 (2017)
17. Ceron, W., de-Lima-Santos, M.F., Quiles, M.G.: Fake news agenda in the era of COVID-19: identifying trends through fact-checking content, Online Soc. Networks Media **21**, 100116 (2021)

Environmental Adaptive Privacy Preserving Contact Tracing System for Respiratory Infectious Diseases

Pengfei Wang[1]([✉]), Xiangyu Su[1], Maxim Jourenko[1], Zixian Jiang[2], Mario Larangeira[1,3], and Keisuke Tanaka[1]

[1] Department of Mathematical and Computing Sciences, School of Computing, Tokyo Institute of Technology, 2 Chome-12-1 Ookayama, Meguro City, Tokyo 152-8550, Japan
{wang.p.ae,su.x.ab}@m.titech.ac.jp
[2] Laboratory of Plasma Membrane and Nuclear Signaling, Graduate School of Biostudies, Kyoto University, Yoshidahonmachi, Kyoto 606-8501, Japan
[3] Input Output HK, Hong Kong, People's Republic of China

Abstract. The COVID-19 pandemic has made the scientific community devise means to implement "contact tracing" mechanisms to mitigate the spread of the infection. The crucial idea is to scan and record close contacts between users using mobile devices, in order to notify persons when their close contact(s) is diagnosed positive. Current contact tracing systems' false-positive rate is too high to be practical as they do not filter Bluetooth scan results outside range of infection. Furthermore current systems neglect airborne transmission other than droplet transmission. Moreover, the ability granted to service providers of the contact tracing systems to access user data violates user privacy. Finally, attackers can modify, remove or fabricate contact records in their devices, which harms the integrity of the system. In this paper, we propose and develop a new contact tracing system which uses environmental factors to filter out results outside estimated effective transmission distance, and also take airborne transmission into consideration. In addition, we implement a rerandomizable signature scheme with blockchain bulletin board to provide confidentiality and integrity. We also evaluate the performance of our theory by implementing our algorithm on mobile devices.

Keywords: Contact-tracing · Traceability · Public key rerandomizable · Blockchain · Epidemiology · Fluid dynamics

1 Introduction

1.1 Background

Infectious diseases have long been one of the most deadly threats to human society. To protect people from being infected by viruses, we will have to track routes of transmission of the viruses and warn potential infectees so that

W. Meng and M. Conti (Eds.): CSS 2021, LNCS 13172, pp. 131–144, 2022.
https://doi.org/10.1007/978-3-030-94029-4_10

suspected patients can quarantine themselves before symptoms grow and avoid spreading the virus.

Starting from Privacy-Preserving Contact Tracing System [1] developed by Apple and Google in April 2020, various organizations or governments around the world developed different kinds of contact tracing systems to combat the COVID-19 pandemic and protect users.

However, current contact tracing systems are not capable of providing either comprehensive security or privacy to users, lacking either precision, privacy or integrity.

To be specific, privacy preserving contact tracing system developed by Apple and Google does not filter results of the scan by Bluetooth Low Energy (BLE) outside the range of transmission. As the range of transmission for droplets is much shorter comparing to the scanning range of BLE (see Sect. 1.2), when all (or even only some) people who got warned try to access hospitals, the yield would be too large for local medical systems to endure.

Also, these systems do not take airborne transmission (different from droplet transmission [15]) into consideration. Droplets may remain in the air for a specific amount of time, while these floating particles may also be infectious during their "lifetime".

Besides, all current contact tracing systems do intend to prevent unauthorized accesses, but they do not categorize service providers as "unauthorized entity", while users probably do.

There has been multiple reports of data leakages of contact tracing database (*e.g.* [17,19,21], etc.), on April 2020, in a joint statement [18] signed by 177 scientists from the U.K., the authors indicated that if there will be a database, it must not allow service providers to de-anonymize its data owners by any means. Blockchain, by its nature, is designed to be pseudonymous, publicly verifiable, non-repudiatable and non-modifiable.

Furthermore, all users of these existing systems can fabricate fake records to flood the system by mimicking authorized packages.

1.2 Related Work

We have evaluated many existing contact tracing systems and concluded that these systems all have some deficiencies (see Table 1). Here, "Demo" means whether a software demonstration exists. "Integrity" means whether someone can fake non-existent records. "Scalability" means whether the system can be deployed in large scale use cases, *e.g.* densely populated scenarios or heavy calculations.

The idea of "privacy preserving contact tracing" is not novel. However, many systems such as Danz et al.'s work [6] only tried to analyze existing schemes and propose a security framework, without trying to establish a practical real application nor solving potential risks, such as linkability of user identifications with contact records, although they did mention the privacy issue caused by linkability. Their contributions limit on summarizing what objectives researchers should seek for a secure contact tracing system and evaluating contemporary works based on these objectives.

Table 1. Comparison with related works

Contact tracing systems			Precision	Demo	Confid.	Integrity	Scalability	Flooding
Real world	Gov.	UK [12]	X	✓	X	X	✓	✓
		JP [14]	X	✓	X	X	✓	✓
		China [9]	X	✓	X	X	N/A	N/A
	Other	GAEN [1]	X	✓	✓	X	✓	✓
Research	Liu et al. [13]		X	△	✓-	X	X	✓
	Canetti et al. [4]		X	X	✓	X	✓	✓
	Kim et al. [11]		X	X	✓	X	X	✓
	An et al. [22]		X	✓	✓	X	X	✓
	Our work		✓	✓	✓	△	✓	✓

More importantly, current works lack precision. Based on our research, there is currently no contact tracing system focusing on reducing false-positive rate of the scanning. For example, the first practical contact tracing system, designed by Apple and Google ("GAEN") [1], uses Bluetooth (BLE) to search for surrounding devices without excluding results outside effect range of droplet transmission. Although the range of BLE are around 50–100 m, distance of droplet transmission is typically between 0–20 m approximately [7]. This means that individuals scanned outside the effective range would be false-positives.

Also, **all** current contact tracing systems ignore the fact that people do not have to contact closely to become "close contacts", because, according to Chen [5], droplets ejected from human body will float for a specific amount of time. These systems do not take time duration of infection into consideration.

Another problem is about privacy preserving. GAEN is a representative model for typical contact tracing. User holds a periodically re-generated temporary key (TEK), to generate other keys using key derivation function, and keys derived from TEK are used to encrypt the payload with AES. After being diagnosed positive, encrypted payload and TEKs will be uploaded, for normal users to trace their contacts. However, for GAEN, its cryptographic design based on AES and random key generation cannot guarantee integrity, *i.e.*, fabricating never-happened meetings or fake identities.

Although all of the contact tracing systems we reviewed emphasize security which by its definition means to prevent unauthorized access, most of them do not clarify who shall be categorized as "unauthorized entities". For example, the contact tracing application designed by N.C.S.C. of the U.K. [12] holds a public key-pair with each user and stores the private key in the server. Hence, the service providers are capable of accessing user data using the private key while preventing unauthorized entities from accessing. However, from users' perspective, service providers and unauthorized entities show no difference as users want hardly anybody but themselves to access their data, while this is not possible with traditional public-key approach.

Systems proposed by Kim et al. [11] and An et al. [22] use functional encryption and homomorphic encryption to circumvent the aforementioned issues from traditional encryptions. However, neither of the primitive is practical in real world. These systems even collect more personally identifiable information than is necessary, *e.g.*, credit card information.

Canetti et al. [4] proposed two contact tracing protocols that achieve different privacy-preserving levels. However, neither of them requires signatures, thus may lack integrity on users' claims. Furthermore, their work sticks with BLE solution, and thus is unable to transmit and authenticate environmental factors safely. We have not found any contact tracing system designed to involve environmental factors in their security frameworks.

Other works, such as Liu et al.'s work [13] would fail to be implemented with heavy traffic, because these works require key exchange protocols and establishing connections. The existing non-interactive contact tracing systems [4,6] would lack integrity, given that users can generate arbitrary pseudonyms without verification. The authors did not provide repository nor screenshot about their implementation as well.

1.3 Our Contribution

Mixing all shortcomings mentioned above, we conclude that improving existing contact tracing approaches is impractical and therefore we decide to design a new contact tracing system to overcome such shortcomings.

The novelty of this work is that, first, we propose a way to use environmental factors to filter scanning results and make the system more precise. Our system is a variant of credential systems and thus it can embed arbitrary environment factors inherently.

Then, according to new requirements (airborne transmission) in contact tracing, our work includes a new contact-tracing scheme, discrete real-time tracking. To solve the dilemma between anonymity and integrity, we implement and modify Su et al.'s work [20], a public key rerandomizable signature scheme with aggregability. The scheme enables users updating their identities to preserve privacy, while keeping integrity with signatures to deny malicious messages, so that users cannot fabricate fake meeting records with other users, undermining integrity of the bulletin board.

We clarify the integration between our tracing system and a bulletin board, which is implemented with a blockchain. We consider several practical issues, such as privacy, scalability and storage overhead. Finally, we developed a demonstration on Android to show the efficiency of our work on mobile devices. Please take note that this research does not limit to the COVID-19, but takes it as an example showing how our system works on pandemics.

2 Environmental Factors

Environmental factors play an important role in contact tracing and we thus evaluate and incorporate a number of epidemiological studies in our system.

There are three ways of transmission for respiratory viruses, as suggested by CDC [15]: contact transmission, droplet transmission, and airborne transmission.

In our system, we neglect contact transmission and assume that all indoor environments are closed, as mobile devices are unable to detect them.

2.1   Overview

Under the assumptions described above, there are three major measurable factors, temperature T, relative humidity RH, and air velocity, and 2 types of transmission, transmission by droplets and by airborne.

This section is represented by box "Discrete Tracing" and "BLE Scanning" in the design diagram (Fig. 2 in Sect. 4).

We have to emphasize that we do not intend to pay attention to the accuracy of epidemiological data and equations mentioned in this section. We believe that such researches are subject to future works of physicists and epidemiologists. Existing contact tracing approaches completely lacked consideration of such fields and our focus is to implement epidemiological studies in contact tracing.

Droplet Transmission. According to Das et al.'s study, the influence of temperature on space-time evolution of droplets' motion, from $0\,°C$ to $40\,°C$, is quite limited (about 10%) [7]. Therefore, we will not take temperature into consideration of droplet transmission.

Instead, sizes of the particles and airflow play significant roles in droplet transmission. Das et al.'s data is based on the Langevin equation [7], Eq. 1.

$$\frac{4}{3}\pi R^3 \rho \frac{d^2 r_i}{dt^2} = -\lambda \frac{dr}{dt} + \xi(t) + F^G \tag{1}$$

In order to measure the size (*i.e.* radius R) of the droplets that determines the mass m (i.e. $4\pi R^3 \rho / 3$), Han et al.'s research provides statistical data of diameter of droplets, measured at 100 ms after the ejection from human body [10].

Together with the Langevin Equation, impact of airflow velocity also needs to be considered, as shown in Eq. 2.

$$v_t = \frac{2R^2(\rho_{droplet} - \rho_{air})g}{9\eta} \tag{2}$$

Das et al. has explained thoroughly about how to solve Eq. 1 and also the effects of airflow velocity in their paper and thus we will not cover these details in this section. Since the authors claim that their experiment data matches calculated assumptions perfectly [7], either of them can be implemented in our system.

We use 0.5 m/s air velocity as the threshold to determine whether the current outdoor environment is ventilated enough, as suggested by NFPA in laboratory ventilation standards [16]. If air velocity is faster than the threshold, transmission radius will be set to 0. This rule also applies to lifetime in airborne transmission as discussed in the following section.

Airborne Transmission. Chen's work provides an important equation in his research as Eq. 3 shows for determining lifetime of the particles. Spalding mass transfer number B is computed from mass fraction $Y_{w,s}$ and $Y_{w,\infty}$. In this equation, temperature has impact on diffusion coefficient D and relative humidity determines the ratio between $Y_{w,s}$ and $Y_{w,\infty}$.

$$t = \frac{\rho_f d_0^2}{8\rho_w D_w(ln(1+B))}, \ where \ B = \frac{Y_{w,s} - Y_{w,\infty}}{1 - Y_{w,s}} \tag{3}$$

Using the equation, data in Chen's research include particles of $1\,\mu m$–$1000\,\mu m$ diameter (d_0), with relative humidity RH from 0% to 100% and temperature at 20 °C and 30 °C. Both data and equation can be implemented in our system to predict lifetime.

Take note that when relative humidity approaches approximately 55.7% [5], lifetime approaches infinity, as $Y_{w,\infty}$ approaches $Y_{w,s}$, shown in Eq. 3.

2.2   Measurement

It is crucial to distinguish indoor environments from outdoor environments, as most countries have some laws or regulations on temperature/humidity in public areas. Using online mapping service, such as Google Maps, it is possible to infer whether the user is staying indoor or outdoor according to the current location and temperature/humidity indoor can be estimated from these laws or regulations.

Besides, outdoor temperature/humidity/airflow is easy to measure—using data from government-held organizations such as the department of meteorology, or from private weather service provider.

3   Contact Tracing

In current contact tracing systems, a contact record will be generated only if two users meet each other, while in case of airborne transmission they do not have to, because of the presence of floating droplets' lifetime, as we have discussed in Sect. 2.1. Therefore, we redesign the scanning mechanism to accomplish our goal.

This section is represented by box "Contact Tracing" in Fig. 2 in Sect. 4.

3.1   Anonymous Discrete Real-Time Tracking

In order for users to detect not only who they are in meeting with, but also whether anyone has been in their current place before, our system need to embed timestamps into records. However, in such case, the definition of "contact records" should be completely different from the current contact tracing systems.

In Fig. 1a, point n denotes locations (previous or current) of the user when discrete records are generated. R_n represents range of transmission, as Sect. 2.1

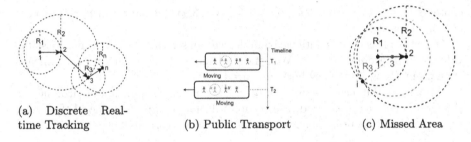

(a) Discrete Real-
time Tracking

(b) Public Transport

(c) Missed Area

Fig. 1. Discrete real-time tracking and exceptions

explained. Arrows merely represent the track of the user in reality under this scenario, without being recorded, as all these points are treated independently by the system to avoid revealing user's track of movement.

For $Record(t_r, t, L, R_n)$ of each point n, record time (t_r), lifetime (t), location-stamp (L) and range (R_n) will be included.

The distance between points of n and $n-1$ should equal to R_{n-1}. In other words, a discrete record will be generated when the user leaves his or her last range of transmission. A new record will be generated periodically if the user is not moving.

Periodically, the system will fetch all discrete points (anonymously) from bulletin board, comparing time, location and range, and determine whether this user should be notified.

3.2 Exceptions

Besides the difference in records, another major difference between our approach and other approaches is that our records denote the absolute position of one user, while other systems record relative position of a pair of users.

Therefore, this will create a problem—when a group of users are moving simultaneously in the same direction and the same speed (*i.e.* public transport, in this scenario), the system will misrecognize users outside the range of transmission as "close contacts" (see Fig. 1b).

Therefore, our solution combines traditional Bluetooth approach with our own discrete real-time approach. When the device is detecting unreasonable speed for walking, it will switch from discrete real-time tracking to traditional Bluetooth-based scanning.

For Bluetooth scanning, to calculate the distance between users, we need to make use of the signal strength (RSSI). Barai et al.'s work [2] indicates that $d = 10^{\frac{MP-RSSI}{10\gamma}}$, where d means distance and MP means measured power.

3.3 Evaluation

The most significant concern about our discrete real-time tracking is that gaps between discrete points (*i.e.* $n = 1$ and $n = 2$) in Fig. 1a may miss some close contacts $(C_3 \cap (\overline{C_1 \cup C_2})$, $R_3 > Distance(n = 3, i))$, as shown in Fig. 1c, where C denotes circle.

If we take Das et al.'s data as example, we can see that the maximum possible value of R_n is around 5 to 6 m. We hence assume that there will not be significant changes in R_n within such short distance.

Also, since the record is uneditable, if there is some significant environmental change after being generated, the record will not be able to respond. For now, we assume that there is no such condition in such short distance.

4   Framework and Threat Models

We have explained every part of contact tracing in previous sections and, in this section, we are going to explain how contact tracing mechanisms are going to be incorporated in our system.

Our work is involved with cryptographic algorithms and is expected to operate on mobile devices, whose performance is comparatively limited comparing to traditional cryptographic scenarios. Therefore, we decide to implement our algorithm in Android application to demonstrate the practicality and performance of our system[1].

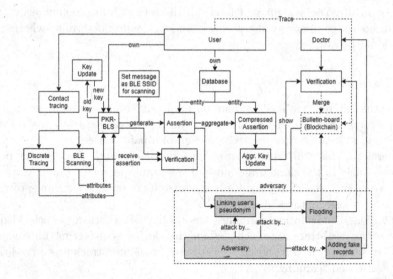

Fig. 2. Overall design of our contact tracing system

---

[1] Anonymous repository of the demonstration due to submission policy: https://anonymous.4open.science/r/PBK_Test-F6C1/.

4.1   Framework

Our framework is illustrated in Fig. 2. A user holds a Database, which has two types of assertions, Assertion[2] and CompressedAssertion. We will explain why we need two types of assertions later in this section. Public keys and signatures needs to be periodically updated to prevent the whole movement track from being exposed.

Database is used for storing Assertions either 1) generated by PKR-BLS using attributes or 2) received from BLE scanning service and verified by PKR-BLS. Assertion(id, pk, attrs, σ, g, g^r) contains auto-generated device-wide unique assertion ID, pseudonym (i.e. public key pk), message (i.e. tracing attributes attrs), signature (i.e. σ), group generator (i.e. g) and contact tracing mode flag (i.e. isBLE).

Database can also be used for storing CompressedAssertion(σ, {id$_i$}) aggregated from Assertions by PKR-BLS. Theoretically, as described in Sect. 5.2, CompressedAssertion should contain an aggregated signature (i.e. SIG.Aggre({σ_i}) $\rightarrow \sigma$), a list of pseudonyms (i.e. {nym$_i$}), a list of attributes (i.e. {attr$_i$}), and a list of group generators (i.e. {g$_i$}).

An infected user will update CompressedAssertion (keys & signatures) using SIG.Update (see Sect. 5.2), and show updated CompressedAssertions to doctors' with attrs and g^{r}s retrieved from Assertions in the Database. Doctors will verify data received and Merge data with the Blockchain.

Additionally, user also holds secret explicit randomness r to generate signature after key updating. If this user use the previous randomness to generate a new signature, key and updated key's relation can be easily inferred if an attacker check if SIG.Vrfy(nym, attrs, σ) = SIG.Vrfy(nym$'$, attrs, σ).

Doctor is only responsible for verifying aggregated signatures, attributes using public keys received from the user. We guarantee that even doctors misbehave and store previous users' record (thus know some pseudonyms), they cannot link those pseudonyms with new-coming users. By the correctness of our PKR-BLS scheme, doctors run VrfyAggre in their verification phase will tell if the aggregate signature is correct and is updated by the same randomness with its corresponding pseudonyms.

During tracing, our system is able to distinguish BLE records from discrete records using isBLE flag, since BLE records and discrete records have different ways to trace. (see Sect. 5.2 and Sect. 3.1 for tracing mechanisms.)

4.2   Threat Model

We assume adversaries being semi-honest, i.e., they follow the protocol definition but intend to collect as much information. We mainly consider cryptographic attack, but argue that our design can also mitigate network attacks, such as DoS attack. However, software or physical attacks are not of our interest, due to

---

[2] We use "Assertion" instead of "Assert" as it is in Sect. 5 because "Assert" is a keyword in Java.

such attacks are inevitable for distributed applications. Deriving from previous works [6], we define the following threat model to our system.

Traceability Correctness. Users can trace all their contacts with diagnosed users by records on the blockchain;
Pseudonym Unlinkability. A user's pseudonyms from different attributes cannot be linked with each other when being used in different close contacts.
Trace Unlinkability. Pseudonyms in records on the blockchain cannot be linked with any other pseudonyms of the same possessor by any user including medical doctors;
Integrity. If the two users have no close contact, regardless one is diagnosed or not, it cannot trigger the other's tracing algorithm. This may only works for contact-tracing but not for discrete-location-tracing.
Flooding Proof. Adversaries should not flood the blockchain;
Privacy Disclosure. Hiding only users' identities may fail to preserve privacy when only a few diagnosed users are in a small area. In such a situation, the user can be identified by the scope of activity, *i.e.*, according to the time and location data in its attributes.

5   Construction from Rerandomizable Signatures

5.1   Overview of Building Block

In this section, we formalize our designs from Sect. 4 and provide concrete cryptographic construction and analysis. Our system works atop a bulletin board, which is implemented with a block-chain, consisting of full-nodes and lightweight nodes. The construction is based on a novel public key rerandomizable BLS signature (PKR-BLS) scheme [20], denoted by SIG. It involves (KeyGen, Sign, Vrfy, Aggre, VrfyAggre, Update, VrfyKP) algorithms. We require **correctness**, *i.e.*, Vrfy, VrfyAggre, and VrfyKP returns 1 on outputs of Sign, Aggre, and Update, respectively; Public keys through the Update algorithm should be indistinguishable (textbfupdate indistinguishable), *i.e.*, for any PPT adversary, given pk and a signing oracle Sign, the probability of distinguishing pk′ ← Update(pk) from pk* ≠ pk is negligible; Finally, we require **unforgeability** and **aggregate unforgeability**, *i.e.*, for any PPT adversary, given a signing oracle Sign, the probability of creating a signature (an aggregate signature) σ on an unqueried message m (of the aggregation list) is negligible. To solve the "rogue key attack" [3] against aggregate unforgeability, we adopt the approach from [3], *i.e.*, reject duplicated messages in VrfyAggre algorithm.

5.2   Protocol Constructions

Our system involves three protocols: key management, contact and tracing.

Key management. Key management is executed by users. The government serves as a black-box trusted agency for key registration.

Setup(1^κ): Take a security parameter κ as input, setup a type-3 bilinear map $e : \mathbb{G}_1 \times \mathbb{G}_2 \to \mathbb{G}_T$ of order q with bit length κ. Output the public parameter $\mathsf{pp} = (q, \mathbb{G}_1, \mathbb{G}_2, \mathbb{G}_T, e, g_1, g_2, \mathsf{H})$;

KeyGen($\mathsf{pp}; r$): Run SIG.KeyGen($\mathsf{pp}; r$) = (pk, sk) and output (pk, sk), where $\mathsf{pk} = (g_2, g_2^r, g_2^{r \cdot \mathsf{sk}})$;

FormNym($\mathsf{pp}, \mathsf{pk}, \mathsf{sk}, \mathsf{attrs}$): On inputs, sample a randomness r' according to time slot index in attrs. SIG.Update(pk, r') $\to$ pk', where $\mathsf{pk}' = (g_2, g_2^{rr'}, g_2^{rr' \cdot \mathsf{sk}})$. Output a pseudonym $\mathsf{nym} = \mathsf{pk}'$.

Contact. Contact is executed among users.

Assert($\mathsf{pp}, \mathsf{nym}, \mathsf{sk}, \mathsf{attrs}; r$): For $\mathsf{nym} = (g_2, g_2^r, g_2^{r \cdot \mathsf{sk}})$, run SIG.Sign($\mathsf{sk}, \mathsf{attrs}; r$) $\to \mathsf{attrs}^{r \cdot \mathsf{sk}}$. Set $\sigma = \mathsf{attrs}^{r \cdot \mathsf{sk}}$ and output $\mathsf{assertion} = (\mathsf{nym}, \mathsf{attrs}, \sigma)$. Notice that Sign algorithm takes in the randomness r corresponds to users' current pseudonym.

Save($\mathsf{pp}, \{\mathsf{assertion}_i\}$): On input a list of assertions, Save first verifies each signatures with SIG.Vrfy($\mathsf{nym}_i, \mathsf{attrs}_i, \sigma_i$). It aggregates the valid signatures with SIG.Aggre($\{\sigma_i\}$) $\to \Pi_{i=1}^n \sigma_i$. Finally, Save sets $\sigma = \Pi_{i=1}^n \sigma_i$ and stores $\mathsf{record} = (\{\mathsf{nym}_i\}, \{\mathsf{attrs}_i\}, \sigma)$.

As a remark, users' attributes should be distinct for PKR-BLS's security. This can be achieved by either more precise environmental factors or embedding randomness into attributes.

Tracing. Tracing is a protocol executed by users and medical doctors. We assume BC as the bulletin board and adopt the light-full node solution, *i.e.*, normal users only store data within a given period (*e.g.*, recent 14 days); while trustworthy volunteers should store the whole chain. The design mitigate the storage overhead for normal user while still preserve the security.

Show($\mathsf{pp}, \mathsf{sk}, \{\mathsf{record}_i\}$): During diagnosis, a user runs SIG.Update($\mathsf{nym}; r$) $\to$ nym' for each $\mathsf{nym} \in \{\mathsf{nym}\}$, and computes $\sigma' = \sigma^r$. Notice that to preserve correctness, the update of pseudonyms and the corresponding aggregate signature requires the same randomness r. The user outputs and shows the updated set $\{\mathsf{record}_i'\} = (\mathsf{nym}_i', \mathsf{attrs}, \sigma')$ to a doctor if diagnosed;

Merge($\mathsf{pp}, \{\mathsf{record}_i\}, \mathsf{BC}$): A medical doctor, on inputs a list of records from a diagnosed user, first runs SIG.VrfyAggre($\{\mathsf{nym}_i\}, \{\mathsf{attrs}_i\}, \sigma$) for every record. It updates with $\mathsf{BC}' = (\mathsf{BC}, (\{\mathsf{nym}_i\}, \{\mathsf{attrs}_i\}))$ for all $\{\mathsf{nym}_i\}, \{\mathsf{attrs}_i\}$ in valid records. Notice the aggregate signature is discarded;

Trace$_{\mathrm{BLE}}$($\mathsf{pp}, \mathsf{sk}, \mathsf{BC}$): A user collects records from the blockchain, runs SIG. VrfyKP($\mathsf{nym}, \mathsf{sk}$) to check if $(g_2^r)^{\mathsf{sk}} = g_2^{r \cdot \mathsf{sk}}$. If the algorithm returns 1, the user is said to have close contacts in the record with $(\{\mathsf{nym}_i\}, \{\mathsf{attrs}_i\})$;

Trace$_{\mathrm{Discrete}}$($\mathsf{pp}, \{\mathsf{record}_i\}, \mathsf{BC}$): A user collects records from the blockchain, checking if $|L_{self} - L_{record}| < R_n$ & $t_r < t$ (see Sect. 2 and 3). Details of comparison can be found in Sect. 3.1.

5.3   Security Analysis

Recall the threat model in Sect. 4.2, traceability correctness is guaranteed by the update correctness of the PKR-BLS scheme, *i.e.*, a user runs SIG.VrfyKP(nym, sk) will output 1 for any its nym $\in$ BC and sk.

For pseudonym unlinkability, in recording phase, users $\mathcal{U}_1$ and $\mathcal{U}_2$ have contact in time slots t, $\mathcal{U}_2$ stores $\mathcal{U}_1$'s assertion signature $\sigma^t \leftarrow$ Sign(nym^t, attrs^t). For any other pseudonyms and signatures of $\mathcal{U}_1$, $\sigma^{t'} \leftarrow$ Sign(nym$^{t'}$, attrs$^{t'}$), PKR-BLS's update indistinguishability guarantees that $\mathcal{U}_2$ cannot distinguish nym^t from nym$^{t'}$ even after learning σ^t and $\sigma^{t'}$.

In the tracing phase, trace unlinkability requires two aspects: For normal users, it holds by PKR-BLS's update indistinguishability; For doctors, users cast a zero-knowledge protocol with the doctor by Show and Merge algorithm.

For integrity, to trigger an honest user's Trace algorithm to output 1, a diagnosed adversary has to create a record involving one of the honest user's assertion. However, without contact, the adversary has no knowledge of user's signature nor can it forge one (by unforgeabilty). Moreover, after the adversary learns ({nym}, {attrs}) from BC and intends to forge a valid record record$_\mathcal{A}$ = ({nym} $\cup$ nym$_\mathcal{A}$, {attrs} $\cup$ attrs$_\mathcal{A}$, $\sigma_\mathcal{A}$), it will fail with significant probability due to the aggregation unforgeability.

6   Experiments and Evaluation

We implemented our system on both Android Virtual Device (AVD) and physical devices (Samsung Galaxy Note 10+ & Google Pixel 3A) to prove its functionality and performance. Furthermore, our application utilizes and modifies the Java Pairing Based Cryptography Library (JPBC) [8] based on our construction.

Table 2. Experiment results (100 Attempts)

Func. frequent	Time consump.	Func. infrequent	Time consump.
Assertion generation	99 ms	Key generation	44 ms
Assertion verification	163 ms	Key update	41 ms
		Save	$f(Amt.Assertions)$

There are roughly three types of functions in our contact tracing system—what needs to be performed frequently, what needs to be performed rarely, and what happens only once (whose performance does not really matter, *e.g.*, Show).

To be specific, KeyGen only needs to be performed once per user generation; Update needs to be performed periodically (*e.g.*, once per hour); Save needs to be performed less frequently than Update (*e.g.*, once per day); Show and its verification only happens during diagnosis. Since time consumption of these function is significantly shorter than time intervals between two launches, as shown in Table 2, we consider our system to be practical regarding these functions.

It needs to be mentioned that time consumption of **Save** depends on the amount of **Assertions** the system trying to aggregate. Their relation is roughly linear based on the result of our experiments.

However, the most time-consuming functions are expected to be performed frequently, *i.e.* assertion generation and assertion verification. Bluetooth scanning scenario requires both generation and verification, while discrete real-time tracking requires only assertion generation.

Consider an extreme case, which a user is tightly surrounded by 100 users. In such case, discrete scenario will take 13.5 s to generate assertions, which is acceptable in our opinion. However, Bluetooth scenario will take 36.3 s to generate and verify assertions, which is significantly slower. However, since Bluetooth scanning only happens in public transports as explained in Sect. 3.2, and it is fast enough for the calculations to be completed before the next stop.

Though, we still expect our system to have notably better performance than current contact tracing systems due to our system's interaction-less nature, especially for discrete tracking, since it does not require assertion reading and verification, shown in Fig. 2.

7   Future Work

There are several unsolved problems which will be discussed in this section.

First, the impact of indoor airflow organization is neglected as we cannot measure it. We hope researchers can do aerodynamic experiments for mobile devices to estimate airflow organization without professional equipment.

Second, we assume that, for outdoor airborne transmission, airflow will rapidly dilute and accelerate the droplets. When droplets is diluted and accelerated, their infectivity will be largely reduced because of lower concentration and faster deactivation by ultraviolet due to faster evaporation caused by acceleration. We hope researchers in the future can validate our assumption.

Third, although users' anonymity is preserved by unlinkability of pseudonyms, the plaintext-assertion of attributes may expose their identities. We thus consider a "blind-asserting" algorithm in the contact protocol (see Sect. 5.2). However, without a proper construction, we list it as future work.

References

1. Apple Inc. and Google LLC: Privacy-preserving contact tracing, April 2020. https://covid19.apple.com/contacttracing
2. Barai, S., Biswas, D., Sau, B.: Estimate distance measurement using NodeMCU ESP8266 based on RSSI technique. In: 2017 IEEE Conference on Antenna Measurements Applications (CAMA), pp. 170–173 (2017)
3. Boneh, D., Gentry, C., Lynn, B., Shacham, H.: Aggregate and verifiably encrypted signatures from bilinear maps. In: Biham, E. (ed.) EUROCRYPT 2003. LNCS, vol. 2656, pp. 416–432. Springer, Heidelberg (2003). https://doi.org/10.1007/3-540-39200-9_26

4. Canetti, R., et al.: Privacy-preserving automated exposure notification. IACR.org, July 2020
5. Chen, L.D.: Effects of ambient temperature and humidity on droplet lifetime - a perspective of exhalation sneeze droplets with COVID-19 virus transmission. Int. J. Hyg. Environ. Health **229**, 113568 (2020)
6. Danz, N., Derwisch, O., Lehmann, A., Puenter, W., Stolle, M., Ziemann, J.: Security and privacy of decentralized cryptographic contact tracing. Cryptology ePrint Archive, Report 2020/1309 (2020)
7. Das, S.K., Alam, J.E., Plumari, S., Greco, V.: Transmission of airborne virus through sneezed and coughed droplets. Phys. Fluids **32**(9), 097102 (2020)
8. De Caro, A., Iovino, V.: JPBC: Java pairing based cryptography. In: Proceedings of the 16th IEEE Symposium on Computers and Communications, ISCC 2011, Kerkyra, Corfu, Greece, June 28–July 1, pp. 850–855. IEEE (2011)
9. Embassy of PRC in UK: Notice on online application for health declaration certificate for non-Chinese nationals, November 2020
10. Han, Z.Y., Weng, W.G., Huang, Q.Y.: Characterizations of particle size distribution of the droplets exhaled by sneeze. J. R. Soc. Interface **10**(88), 20130560 (2013)
11. Kim, W., Lee, H., Chung, Y.D.: Safe contact tracing for COVID-19: a method without privacy breach using functional encryption techniques based-on spatio-temporal trajectory data. PLOS ONE **15**(12), e0242758 (2020). https://doi.org/10.1371/journal.pone.0242758
12. Levy, I.: High level privacy and security design for NHS COVID-19 contact tracing app. Technical report, National Cyber Security Centre, United Kingdom, May 2020
13. Liu, J.K., et al.: Privacy-preserving COVID-19 contact tracing app: a zero-knowledge proof approach. Cryptology ePrint Archive, Report 2020/528 (2020)
14. Ministry of Health: Covid-19 contact-confirming application. Technical report, Labour and Welfare of Japan, Tokyo, Japan, December 2020
15. National Center for Immunization and Respiratory Diseases: Scientific brief: SARS-COV-2 and potential airborne transmission. Technical report, Centers for Disease Control and Prevention, Atlanta, Georgia, United States, October 2020
16. National Fire and Protection Agency: NFPA-45 standard on fire protection for laboratories using chemicals, section 6-4.5 (2019)
17. Shen, X.: Personal information collected to fight Covid-19 is being spread online in China. South China Morning Post, May 2020. Accessed Feb 2021
18. Signees of the Joint Statement: Joint statement, April 2020. https://drive.google.com/file/d/1uB4LcQHMVP-oLzIIHA9SjKj1uMd3erGu/view
19. Starks, T.: Early Covid-19 tracking apps easy prey for hackers, and it might get worse before it gets better. Politico, July 2020. Accessed Feb 2021
20. Su, X., Wang, P., Jourenko, M., Larangeira, M., Tanaka, K.: Contact tracing from BLS signature with updatable public keys. In: SCIS (2021)
21. Whittaker, Z.: Fearing coronavirus, a Michigan college is tracking its students with a flawed app. TechCrunch, August 2020. Accessed Feb 2021
22. An, Y., et al.: Privacy-oriented technique for COVID-19 contact tracing (PROTECT) using homomorphic encryption: design and development study. J. Med. Internet Res. **23**(7), e26371 (2021). https://doi.org/10.2196/26371

A Privacy-Preserving Logistics Information System with Traceability

Quanru Chen[1], Jinguang Han[2(✉)], Jiguo Li[3,4], Liquan Chen[5], and Song Li[1]

[1] College of Information Engineering, Nanjing University of Finance
and Economics, Nanjing, China
lisong@nufe.edu.cn
[2] Jiangsu Provincial Key Laboratory of E-Business, Nanjing University of Finance
and Economics, Nanjing, China
9120131003@nufe.edu.cn
[3] College of Computer and Cyber Security, Fujian Normal University, Fuzhou, China
[4] Fujian Provincial Key Laboratory of Network Security
and Cryptology, Fuzhou, China
[5] School of Cyber Science and Engineering, Southeast University, Nanjing, China

Abstract. Logistics Information System (LIS) is an interactive system
that provides information for logistics managers to monitor and track
logistics business. In recent years, with the rise of online shopping, LIS
is becoming increasingly important. However, since the lack of effective
protection of personal information, privacy protection issue has become
the most problem concerned by users. Some data breach events in LIS
released users' personal information, including address, phone number,
transaction details, etc. In this paper, to protect users' privacy in LIS, a
privacy-preserving LIS with traceability (PPLIST) is proposed by combining multi-signature with pseudonym. In our PPLIST scheme, to protect privacy, each user can generate and use different pseudonyms in
different logistics services. The processing of one logistics is recorded
and unforgeable. Additionally, if the logistics information is abnormal, a
trace party can de-anonymize users, and find their real identities. Therefore, our PPLIST efficiently balances the relationship between privacy
and traceability.

Keywords: Privacy protection · Multi-signature · Pseudonym ·
Traceability · Logistics information system

1 Introduction

In recent years, with the rapid development of e-commerce, online shopping has
become a popular trend. Online shopping is an interactive activity between a
buyer and a seller, where after completing an order by a buyer, the product is
delivered via a logistics system [27]. Logistics system helps to reduce product
cost and save shopping time.

© Springer Nature Switzerland AG 2022
W. Meng and M. Conti (Eds.): CSS 2021, LNCS 13172, pp. 145–163, 2022.
https://doi.org/10.1007/978-3-030-94029-4_11

Unfortunately, the current LISs [34] cannot effectively protect users' privacy information. Users' personal information is clearly visible on the express bill and the LIS database [49]. Some data breaches in LISs released users' personal information, including addresses, phone numbers, transaction details, etc. If a user's personal information is leaked and maliciously collected, she may be at high risk of identity forgery and property fraud, in addition to the risk of being harassed by spam messages. Therefore, it is interesting and important to consider the privacy issues in LISs.

Furthermore, since a product is delivered by multiple logistics stations, it is important to record the whole logistics process and make the process unforgeable. Additionally, to prevent users from conducting illegal transactions, users can be de-anonymized [36,43] and traced.

In this paper, we propose a privacy-preserving logistics information system with traceability (PPLIST). Compared with the existing LISs, our scheme has the following advantages:

1) Users can anonymously use the logistics services in our PPLIST scheme. Users generate and use different pseudonyms in different logistics services. Even the internal staff of a logistics company cannot directly obtain the information of users' identities, our PPLIST scheme effectively protects users' personal information.
2) In the case that the identity of a user needs to be released, a trace party can de-anonymize a user and find his identity. This property prevents users from conducting illegal logistics via a logistics system.
3) Our PPLIST scheme is efficient. Multi-signature is applied to record the delivery process and reduces the storage space.

Contributions: Our main contributions in this paper are summarised as follows: 1) The definition and security model of our PPLIST scheme are formalised; 2) A PPLIST scheme is formally constructed; 3) The security of our PPLIST scheme is formally reduced to well-known complexity assumptions; 4) Our PPLIST scheme is implemented and evaluated.

1.1 Related Work

In this subsection, we introduce the work which is related to our PPLST scheme, including LIS, privacy protection in LIS, multi-signature and pseudonym.

Logistics Information System. LIS is a subsystem and the nerve center of logistics systems. As the control center of the whole logistics activities, LIS has many functions. The main functions of LIS are as follows: collect, store, transmit, process, maintain and output logistics information; provide strategic decision support for logistics managers; improve the efficiency of logistics operations [29].

Bardi et al. [6] pointed out that the choice of LIS directly affected the logistics cost and customer-service level. Lai et al. [28] showed that LIS is very important for a company to manage product inventory and predict the trend of customers' online shopping. In addition, Ngai et al. [38] claimed that LIS is an information system that can promote a good communication between the companies and the customers. An LIS adoption model was proposed in [38] to examine the relationship among organizational environment, perceived benefits and perceived barriers of LIS adoption. In [16], Closs and Xu argued that the important source of enterprise competitive advantages was logistics information technology. Their research showed that companies with advanced logistics information technology and LIS performed better than other companies.

LISs have been proposed and applied into various application scenarios [1,2]. Amazon [1] is one of the first companies to provide e-commerce services. Amazon has a logistics system, which realizes the organization and operation of the whole logistics activities. Amazon has also added the special technology, One-Click [15], in their LIS, which can automatically store the information of customers. Therefore, customers do not input their person information in each shopping. In addition, Amazon's LIS has the following functions [17]: order confirmation in time, smooth logistics process, accurate inventory information and optional logistics methods, etc. Amazon has become a business to consumer (B2C) e-commerce [26] company.

Taobao [2] is a consumer to consumer (C2C) e-commerce [52] platform. Taobao entrusts all logistics activities to a third party logistics company, but takes a series of measures to ensure the security of logistics activities. For instance, Taobao implements the network real-name system (NRS) [44] in their LIS, and has set up a special customer-service department to solve products logistics problems. Besides, Taobao has the functions of timely confirmation of orders and delivery within the specified time.

Privacy Protection in LIS. Although the LIS of e-commerce platform brings convenience to people's life, it also brings great challenges to privacy protection. LIS stores a large number of users' personal information. Once the information is leaked, it will result in serious threaten to the life and safety of users. Some privacy protection methods in LIS have been proposed, such as [18,30,33,42,46, 48,50]. We compare our scheme with these systems in Table 1.

Léauté et al. [33] proposed a scheme to ensure the privacy of users while minimizing the cost of logistics operation. The scheme formalizes the problem as a Distributed Constraint Optimization Problem (DCOP) [20], and combines various techniques of cryptography. But the disadvantage of this scheme [33] is that the anonymization of users is not considered. In [50], Frank et al. proposed a set of protocols for tracking logistics information, which is a light-weight privacy protection mechanism.

To solve the problem of privacy leakage caused by stolen express order number, Wei et al. [48] proposed a k-anonymous model to protect logistics information. However, the method only protects a part of users' personal information, because the names and telephone numbers of receivers are directly printed on the express bills for delivery.

Table 1. The comparison between our scheme and related schemes.

Systems	Anonymity	Traceability	Security proof
Gao et al. [18]	×	×	✓
Léauté et al. [33]	×	×	✓
Qi et al. [42]	×	✓	×
Liu et al. [30]	✓	×	✓
Laslo et al. [46]	×	✓	×
Wei et al. [48]	✓	×	×
Frank et al. [50]	×	✓	×
Our PPLIST	✓	✓	✓

To improve the security of [48], Qi et al. [42] proposed a new logistics management scheme based on encrypted QR code [46]. After a courier scans the encrypted QR code by using an APP, the logistics information of products in the database is automatically updated through GPRS or Wi-Fi. The APP provides an optimal delivery route for couriers. However, the problem of [42] is that users' personal information is still visible to the internal staff of express companies. In addition, Laslo et al. [46] proposed a traceable LIS based on QR code. However, this scheme does not consider privacy information protection.

Furthermore, Gao et al. [18] proposed a secure LIS, named LIP-PA, which can protect the logistics process information between different logistics stations, but the protection of users' personal information is not considered well. Hence, the privcay of users in LISs [18,42,46] was not fully considered.

Liu et al. [30] designed an LIS based on the Near Field Communication (NFC) [51] technology. In [30], users' personal information was hidden in tags, and only authorized people can access information. However, because of the limitation of computation power, the scheme cannot perform complex encryption and decryption processes.

In summary, above schemes addressed the privacy issues in LIS, but these scheme did not consider the track of delievery process and the trace of illegal users. However, these are important issues in LISs. Therefore, to solve these problems, we propose a new privacy-preserving LIS called PPLIST.

Multi-signature. Multi-signature, also called multi-digital signature, is an important branch of digital signature. Multi-signature is suitable to the case where multiple users sign on a message, and a verifier is convinced that each user participated in the signing [8].

Itakura [24] first proposed the concept of multi-signature, and proposed a multi-signature scheme with fixed number of signatures. Then, many multi-signature schemes were proposed [7,23,35,39–41]. The multi-signature generation time of schemes [24,41] is linear with the number of signers. Okamoto et al. [39] proposed a multi-signature scheme, but it, like scheme [40], only allows each signer in a group to sign messages. It's inflexible. Furthermore, Ohta and Okamoto [40] formlized the security model of multi-signature. However, this scheme did not consider the security of the key generation process, so its security is not strong. Based on [40], Micali et al. [35] proposed a formal and strong security model for multi-signature. Bellare and Neven [7] proposed a new scheme and proved its secure in the plain public-key model. This scheme improved the efficiency of previous multi-signature schemes.

Since it enables multiple signers to collabratively sign on a message, multi-signature has been used into various application scenarios, such as [4,9,37,47]. Shacham [31] proposed a sequential aggregate multi-signature scheme. The scheme computed the final multi-signature by sequentially aggregating the signatures from multiple signers. However, the data transmission of [31] is large. To solve this problem, Neven [37] presented a new sequential aggregate multi-signature scheme based on [31]. The scheme of [37] reduces signing and verification costs effectively.

Tiwari et al. [47] proposed a secure multi-proxy multi-signature scheme. It does not need paring operations, and reduces the running time. The scheme is aslo secure against the attack of selected messages. However, Asaar et al. [4] found the scheme in [47] is insecure, and proposed an identity-based multi-proxy and multi-signature scheme without pairings. The security of this scheme was reduced to the RSA assumption in the random oracle model by using the Forking Lemma technique [5].

Recently, Dan et al. [9] proposed a new multi-signature scheme. Signature compression and public-key aggregation were used in the scheme. Therefore, when a group of singers signed a message, the verifier only needs to verify the final aggregate signature. The advantage of this scheme is that the size of the final aggregate signature is constant and independent of the number of signers. Furthermore, this scheme is secure against rogue-key attacks. When constructing our PPLIST, we apply the scheme [9] to record the whole logistics process and reduces the storage cost.

Pseudonym. Pseudonym is a method that allows users to interact anonymously with other organizations. Because pseudonym is unlinkable, it can effectively protect the information of a user's identity [45] among multiple authentications. The common pseudonym generation techniques are as follows [14]: 1) Encryption with public key; 2) Hash function; 3) Keyed-hash function with stored key; 4) Tokenization.

Chaum [10] found that pseudonym enables users to work anonymously with multiple organizations, and users can use different pseudonyms in different organizations. Because of the unlinkability of pseudonym, no organization can link a user's pseudonymes to her identities. Later, Chaum and Evertse [11] presented a pseudonym model scheme based on RSA. However, the scheme needs a trusted center to complete the signing and transfer of all users' credentials.

To reduce the trust on the trusted center, Chen [12] proposed a scheme based on discrete logarithm. The scheme also needs a trusted center, but the trusted center is only required for pseudonym verification. Although Chen's scheme is less dependent on the trusted center than the scheme [11], the trusted center was still required.

In order to enable users to have the initiative in the pseudonym system, Lysyanskaya et al. [32] proposed a new scheme. In this scheme, a user's master secret key was introduced. If the master secret keys are different, the information of users' identities must be different. In addition, the pseudonym certificate submitted by a user to an organization only corresponds to the user's master public key and does not disclose the information of his master secret key.

Pseudonym has been applied in some schemes [21,25] to protect users' privacy. To reduce the communication cost of traditional pseudonym systems in Internet of Vehicles, Kang et al. [25] proposed a privacy-preserved pseudonym scheme. In this scheme, the network edge resources were used for effective management, and the communication cost was effectively reduced.

In [21], Han et al. proposed an anonymous single sign-on (ASSO) scheme. In this scheme, pseudonym was applied to protect users' identities. A user uses his secret key to generate different pseudonyms, and obtains a ticket from a ticket issuer anonymously without releasing anything about her real identity. Furthermore, a user can use different pseudonyms to buy different tickets and the ticket issuer cannot know whether two tickets are for the same user or two different users. In our PPLIST scheme, to protect users' privacy, we apply the pseudonym developed in [21] to enable users to use logistics services anonymously and unlinkably.

1.2 Paper Organisation

The remainder of this paper is organised as follows. Section 2 presents the preliminaries used in our scheme, and describes the formal definition and security model of our PPLIST scheme. Section 3 provides the construction of our scheme. The security proof and implementation of our scheme are presented in Sect. 4 and Sect. 5, respectively. Finally, Sect. 6 concludes this paper.

2 Preliminaries

In this section, the preliminaries used throughout this paper are introduced, including bilinear groups, complexity assumptions, formal definition and security model. Table 2 summaries the notations used in this paper.

The framework of our PPLIST is presented in Fig. 1. The system first generates the public parameters PUB. Then, each entity (e.g. logistics station, user and the trace party) generates its secret-public key pair. Prior to ordering a service, the user U generates a pseudonym by using his secret key. The system determines the delivery path, and then generates the aggregate public key of the selected logistics stations. After that, each selected logistics station S_i generates its single signature Sig_i on the product information, pseudonym and aggregated public key, and then passes it to the next selected logistics station. Finally, the

Table 2. Notation summary

Notation	Explanation	Notation	Explanation
1^l	A security parameter	$Pseudonym$	The pseudonym of U
S_i	The i-th logistics station	PUB	Public parameters
U	User	PPT	Probable polynomial-time
T	The trace party	$\mathscr{B}(1^l)$	A bilinear group generator
YA	The aggregation of AgY	$x \xleftarrow{R} X$	x is randomly selected from X
AgY	A set of selected public keys	H_1, H_2, H_3	Cryptographic hash functions
σ	The aggregation of signatures	Sig_i	The i-th single signature
π	The proof of user's ownership	I	A set consisting of the indexes of selected logistics stations
d	The number of elements in I		
q	A prime number		

last selected logistic station generates its signature and the aggregate signature σ. To obtain the product, the user needs to prove that he is the owner by generating a proof of the knowledge included in the pseudonym. The user can verify whether the product is delivered correctly by checking the aggregate signature σ. In the case that the identity of a user needs to be traced, the trace party T can use his secret key to de-anonymous the pseudonym, and find the user's identity.

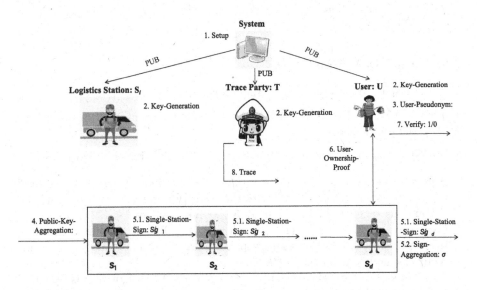

Fig. 1. The framework of our PPLIST scheme

2.1   Bilinear Groups

Let G_1, G_2, G_ι be cyclic groups with prime order q. A map $e : G_1 \times G_2 \to G_\iota$ is a bilinear map if it satisfies the following properties: **(1) Bilinearity:** For all $g \in G_1$, $h \in G_2$, $a, b \in Z_q$, $e(g^a, h^b) = e(g^b, h^a) = e(g, h)^{ab}$; **(2) Non-degeneration:** For all $g \in G_1$, $h \in G_2$, $e(g, h) \neq 1_\iota$, where 1_ι is the identity element in G_ι; **(3) Computability:** For all $g \in G_1$, $h \in G_2$, there exists an efficient algorithm to compute $e(g, h)$.

Let $\mathscr{B}(1^l) \to (e, q, G_1, G_2, G_\iota)$ be a bilinear group generator which takes as input a security parameter 1^l and outputs a bilinear group $(e, q, G_1, G_2, G_\iota)$.

A function $\epsilon(x)$ is negligible if for any $k \in N$, there exist a $z \in N$ such that $\epsilon(x) < \frac{1}{x^z}$ when $x > z$.

2.2   Complexity Assumptions

Definition 1 (Computational Diffie-Hellman (CDH) Assumption [22]). *Let $\mathscr{B}(1^l) \to (e, q, G_1, G_2, G_\iota)$, and g_1, g_2 be generators of G_1, G_2, respectively. Suppose that $\alpha, \beta \xleftarrow{R} Z_q$. Given a triple $(g_1^\alpha, g_1^\beta, g_2^\beta)$, we say that the CDH assumption holds on $(e, q, G_1, G_2, G_\iota)$ if all PPT adversaries $\mathcal{A}$ can output $g_1^{\alpha\beta}$ with a negligible advantage, namely $Adv_{\mathcal{A}}^{CDH} = PR[\mathcal{A}(g_1^\alpha, g_1^\beta, g_2^\beta) \to g_1^{\alpha\beta}] \leq \epsilon(l)$.*

Definition 2 (Discrete Logarithm (DL) Assumption [19]). *Let G be a cyclic group with prime order q, and g be a generator of G. Given $Y \in G$, we say that the DL assumption holds on G if all PPT adversaries can output a number $x \in Z_q$ such that $Y = g^x$ with a negligible advantage, namely $Adv_{\mathcal{A}}^{DL} = PR[Y = g^x | \mathcal{A}(q, g, G, Y) \to x] \leq \epsilon(l)$.*

2.3   Formal Definition

A PPLIST scheme is formalized by the following eight algorithms:

$Setup(1^l) \to PUB$. The algorithm takes the security parameters 1^l as input and outputs the public parameters PUB.

$Key - Generation$. This algorithm consists of the following sub-algorithms:

1) $Station - Key - Generation(1^l) \to (SK_{S_i}, PK_{S_i})$. This algorithm is executed by each logistics station S_i. S_i takes the security parameters 1^l as input, and outputs his secret-public key pair (SK_{S_i}, PK_{S_i}), where $i = 1, 2, 3, \cdots, n$.
2) $User - Key - Generation(1^l) \to (SK_U, PK_U)$. This algorithm is executed by a user U. U takes the security parameters 1^l as input, and outputs his secret-public key pair (SK_U, PK_U).
3) $Trace - Key - Generation(1^l) \to (SK_T, PK_T)$. This algorithm is executed by a trace party T. T takes the security parameters 1^l as input, and outputs his secret-public key pair (SK_T, PK_T).

$User - Pseudonym(PUB, SK_U, PK_T) \rightarrow Pseudonym$. This algorithm is executed by U. U takes as input his secret key SK_U, the public key PK_T of the trace party and the public parameters PUB, and outputs a pseudoym $Pseudonym$.

$Public - Key - Aggregation(PUB, PK_{S_{k_1}}, PK_{S_{k_2}} \cdots, PK_{S_{k_d}}) \rightarrow YA$. Let I be a set which consists of the indexes of some selected logistics stations. This algorithm takes as input the public parameters PUB and the public keys $PK_{S_{k_1}}, PK_{S_{k_2}}, \cdots, PK_{S_{k_d}}$ of selected logistics stations, and outputs the aggregate public key YA.

$Sign$. This algorithm consists of the following sub-algorithms:

1) $Single - Station - Sign(PUB, SK_{S_{k_i}}, Pseudonym, YA, m) \rightarrow Sig_i$. This algorithm is executed by each selected logistics station S_{k_i}. S_{k_i} takes as input its secret key $SK_{S_{k_i}}$, the aggregated public key YA, product information m and the public parameters PUB, and outputs a signature Sig_i, where $i = 1, 2, 3, \cdots, d$.

2) $Sign - Aggregation(PUB, Sig_1, Sig_2, \cdots, Sig_d) \rightarrow \sigma$. This algorithm takes as input the public parameters PUB and signatures Sig_i, and outputs a final signature σ.

$User - Ownership - Verify(PUB, SK_U, PK_T, Pseudonym) \leftrightarrow S_i(PUB) \rightarrow (\pi, 1/0)$. This algorithm is executed between S_i and U.

1) U takes as input his secret key SK_U, the $T's$ public key PK_T, his pseudonym $Pseudonym$ and the public parameters PUB, and outputs a proof π.
2) The verifier takes as input the public parameters PUB, and outputs 1 if the proof π is valid; otherwise, it outputs 0 to show the proof is invalid.

$Verify(PUB, \sigma, Pseudonym, YA, m) \rightarrow 1/0$. This algorithm takes as input the public parameters PUB, the final signature σ, the pseudonym $Pseudonym$, the aggregate public key YA and product information m, and outputs 1 if the signature σ is valid; otherwise, it outputs 0 to show it is an invalid signature.

$Trace(PUB, \sigma, SK_T, Pseudonym, YA, m) \rightarrow PK_U/\perp$. This algorithm is executed by T. T takes as input his secret key SK_T, the pseudonym $Pseudonym$, the aggregate public key YA, the final signature σ, product information m and the public parameters PUB, and outputs $U's$ public key PK_U if the signature σ is valid; otherwise, it outputs $\perp$ to show failure.

2.4   Security Requirements

The security model of our scheme is defined by the following two games.

Unforgeability. This is used to define the unforgeability of signatures, namely even if users, the trace party and the other stations collude, they cannot forge a valid signature on behalf of the selected logistics stations. This game is exectued between a challenger C and a forger $\mathcal{F}$.

Setup. $\mathcal{C}$ runs $Setup(1^l) \to PUB$ and sends PUB to $\mathcal{F}$.

Key-Generation Query.

1) $\mathcal{F}$ asks the public key of stations. $\mathcal{C}$ runs $Station - Key - Generation(1^l) \to (SK_{S_i}, PK_{S_i})$ and sends the station's public key PK_{S_i} to $\mathcal{F}$.
2) When $\mathcal{F}$ asks a urse's secret-public key pair, $\mathcal{C}$ runs $User - Key - Generation(1^l) \to (SK_U, PK_U)$ and sends (SK_U, PK_U) to $\mathcal{F}$. Let GPU be a set of users' public key.
3) When $\mathcal{F}$ asks the secret-public key pair of the trace party, $\mathcal{C}$ runs $Trace - Key - Generation(1^l) \to (SK_T, PK_T)$ and sends (SK_T, PK_T) to $\mathcal{F}$.

User-Pseudonym Query. $\mathcal{F}$ submits a SK_U and the public key PK_T of the trace party, $\mathcal{C}$ runs $User - Pseudonym(PUB, SK_U, PK_T) \to Pseudonym$ and sends $Pseudonym$ to $\mathcal{F}$. Let UPQ be a set of pseudonyms of users.

Public-Key-Aggregation Query. Let I be a set which consists of the indexes of some selected logistics stations and let d be the number of elements in the set I. $\mathcal{F}$ submits a group of selected stations' public keys. $\mathcal{C}$ runs $Public - Key - Aggregation(PUB, PK_{S_{k_i}}) \to YA$, where $i = 1, 2, \cdots, d$. $\mathcal{C}$ returns YA to $\mathcal{F}$.

Sign Query. $\mathcal{F}$ adaptively submits selected station's secret key $SK_{S_{k_i}}$, the aggregation of public key YA, and $U's$ pseudonym $Pseudonym$ and the product information m to ask for a single signature Sig_i up to ϱ times.

Output. $\mathcal{F}$ outputs a forged signature Sig_i', a final signature σ', $U's$ pseudonym $Pseudonym$ and the product information m', the public keys of selected logistics stations AgY and the aggregated public keys YA'. $\mathcal{F}$ wins the game if $PK_{S_i} \in AgY$, $\mathcal{F}$ has not conducted signature query on the message m', and $Verify(PUB, \sigma', Pseudonym, YA', m') = 1$.

Definition 3. *A privacy-preserving logistics information system with traceability is $(\varrho, \epsilon(l))$ unforgeable if all probabilistic polynomial-time (PPT) forger $\mathcal{F}$ who makes ϱ signature queries can only win the above game with a negligible advantage, namely*

$$Adv = Pr\left[Verify(PUB, \sigma', Pseudonym, YA', m') = 1\right] \leq \epsilon(l) \qquad (1)$$

Traceability. This is used to formalise the traceability of our scheme, namely an attacker $\mathcal{A}$ cannot frame a user who did not use the logistics services. We suppose that at least one station is honest. This game is exectued between a challenger $\mathcal{C}$ and an attacker $\mathcal{A}$.

Setup. $\mathcal{C}$ runs $Setup(1^l) \to PUB$ and sends PUB to $\mathcal{A}$.

Key-Generation Query

1) $\mathcal{A}$ can ask for the public key of each station. $\mathcal{C}$ runs $Station - Key - Generation(1^l) \rightarrow (SK_{S_i}, PK_{S_i})$ and sends the station's public key PK_{S_i} to $\mathcal{F}$.
2) When $\mathcal{A}$ asks a urse's secret-public key pair, $\mathcal{C}$ runs $User - Key - Generation(1^l) \rightarrow (SK_U, PK_U)$. Let the secret-public key pair of U^* is (SK_{U^*}, PK_{U^*}). $\mathcal{C}$ sends other users' secret-public key pair (SK_U, PK_U) and PK_{U^*} to $\mathcal{A}$. Let GPU be a set consisting of users's public keys.
3) When $\mathcal{A}$ asks the secret-public key pair of the trace party, $\mathcal{C}$ runs $Trace - Key - Generation(1^l) \rightarrow (SK_T, PK_T)$ and sends (SK_T, PK_T) to $\mathcal{A}$.

User-Pseudonym Query. $\mathcal{A}$ submits a user's SK_U and the public key PK_T of the trace party, $\mathcal{C}$ runs $User - Pseudonym(PUB, SK_U, PK_T) \rightarrow Pseudonym$ and sends $Pseudonym$ to $\mathcal{A}$. Let UPQ be a set of pseudonyms of users.

Public-Key-Aggregation Query. Let I be a set which consists of the indexes of some selected logistics stations and let d be the number of elements in the set I. $\mathcal{A}$ submits a group of selected stations' public keys. $\mathcal{C}$ runs $Public - Key - Aggregation(PUB, PK_{S_{k_i}}) \rightarrow YA$, where $i = i, 2, \cdots, d$. $\mathcal{C}$ returns YA to $\mathcal{A}$.

Sign Query. $\mathcal{A}$ adaptively submits a selected station's secret key $SK_{S_{k_i}}$, the aggregation of public key YA, and $U's$ pseudonym $Pseudonym$ and the product information m to ask for a single signature Sig_i up to ϱ times.

Output. $\mathcal{A}$ outputs a tuple $(\sigma', Pseudonym', YA', m')$. $\mathcal{A}$ wins the game if $Trace(PUB, \sigma', SK_T, Pseudonym', YA', m') \rightarrow PK'_{U^*}$ with $PK'_{U^*} \notin GPU$ or $PK'_{U^*} \neq PK_{U^*} \in GPU$.

Definition 4. *A privacy-preserving logistics information system with traceability is $(\varrho, \epsilon(l))$ traceable if all probabilistic polynomial-time (PPT) adversary $\mathcal{F}$ who makes ϱ signature queries can only win the above game with a negligible advantage, namely*

$$Adv = Pr\left[\begin{array}{c} PK'_{U^*} \notin GPU \ or \\ PK'_{U^*} \neq PK_{U^*} \in GPU \end{array} \middle| \begin{array}{c} Trusted - Party - Trace \\ (PUB, \sigma', SK_T, Pseudonym', \\ YA', m') \rightarrow PK'_{U^*} \end{array} \right] \leq \epsilon(l)$$

$$(2)$$

3 Construction of Our Scheme

In this section, we review the construction of our scheme. We firstly present a high-level overview, and then describe the formal construction of our scheme.

3.1 High-Level Overview

The high-level overview of our scheme is as follows.

Setup. The system generates the corresponding public parameters PUB.

Key-Generation. Suppose that there are n logistics stations. Each S_i, U and T generate their secret-public key pairs (x_{s_i}, Y_{s_i}), (x_u, Y_u) and (x_t, Y_t), where $i = 1, 2, 3, \cdots, n$.

User-Pseudonym. In order to protect privacy in a delivery process, U generates a pseudonym $Pseudonym$ by using his secret key SK_U and $T's$ public key PK_T.

Public-Key-Aggregation. According to product information, the system determines the logstics process by selecting a set of logistcs stations $S_{k_1}, S_{k_2}, \cdots, S_{k_d}$. Let $AgY = \{Y_{S_{k_1}}, Y_{S_{k_2}}, \cdots, Y_{S_{k_d}}\}$ be a set consisting of the public keys of the selected logistics stations. For each service, a table $Table$ is built to record its delivery information. The system uses the public key of each S_{k_i} and the set AgY to generate h_i to resist the rogue key attacks, where $i = 1, 2, 3, \cdots, d$. Then, the system generates the aggregated public key YA.

Sign. Each selected logistics station S_{k_i} uses his secret key $x_{s_{k_i}}$ to generate a signature Sig_i on $U's$ pseudonym $Pseudonym$ and the product information m, and sends Sig_i to the next logistics station. Finally, the last logistic station S_{k_d} uses his secret key x_{k_d} to generate a single signature Sig_d on $U's$ pseudonym $Pseudonym$ and the produce information m, and compute the aggregate signature $\sigma = \prod_{i=1}^{d} Sig_i$. S_{k_d} also adds σ into the table $Table$.

User-Ownership-Verify. When U proves to the last logistic station S_{k_d} that he is the owner of the product, he uses his secret key to prove that his secret key x_u is included in the pseudonym $Pseudonym$ by executing a zero-knowledge proof with S_{k_d}. If the proof is correct, U is the owner of the product; otherwise, he is not the owner of the product.

Verify. When U receives a product, he verifies whether the product was delivered correctly by checking the validity of the aggregate signature σ. If it is, the product is delivered correctly; otherwise, there are some problems in the delivery process.

Trace. Given $(\sigma, Pseudonym, AgY, m)$, in the case that a user needs to be de-anonymized, the trace party T first checks whether the signature is correct or not. If it is incorrect, T aborts; otherwise, T uses his secret key x_t to de-anonymize the Pseudonym and get U's public key Y_u.

3.2   Formal Construction

The formal construction of our PPLIST scheme is formalised by the following eight algorithms:

Setup. The system runs $\mathscr{B}(1^l) \rightarrow (e, q, G_1, G_2, G_\iota, g_1, g_2)$ with $e : G_1 \times G_2 \rightarrow G_\iota$. Let g_1 be a generator of G_1 and g_2 be a generator of G_2. Suppose that $H_1 : \{0,1\}^* \rightarrow G_1, H_2 : \{0,1\}^* \rightarrow Z_q$ and $H_3 : \{0,1\}^* \rightarrow Z_q$ are cryptographic hash functions. The public parameters are $PUB = (e, q, G_1, G_2, G_\iota, g_1, g_2, H_1, H_2, H_3)$.

Key-Generation

1) *Station − Key − Generation.* Each logistics station S_i selects $x_{s_i} \xleftarrow{R} Z_q$ and computes $Y_{s_i} = g_2^{x_{s_i}}$. The secret-public key pair of S_i is (x_{s_i}, Y_{s_i}), where $i = 1, 2, 3, \cdots, n$.

2) *User − Key − Generation.* Each U selects $x_u \xleftarrow{R} Z_q$ and computes $Y_u = g_2^{x_u}$. The secret-public key pair of U is (x_u, Y_u).

3) *Trace − Key − Generation.* T selects $x_t \xleftarrow{R} Z_q$ and computes $Y_t = g_2^{x_t}$. The secret-public key pair of T is (x_t, Y_t).

User-Pseudonym. To generate a pseudonym for a product information m, U firstly computes $k = H_3(x_u \parallel m)$ and then computes $C_1 = g_2^k, C_2 = Y_t^k \cdot g_2^{x_u}$. The pseudonym is $Pseudonym = (C_1, C_2)$ which is the ElGamal encryption of the user U's public key under the trace party T's pubic key.

Public-Key-Aggregation. Let $AgY = \{Y_{S_{k_1}}, Y_{S_{k_2}}, \cdots, Y_{S_{k_d}}\}$ be a set consisting of the public keys of the logistics stations which will deliver the product to the user. The system firstly computes $h_i = H_2(Y_{S_{k_i}} \parallel AgY)$, and then computes $YA = \prod_{i=1}^{d} Y_{S_{k_i}}^{h_i}$. Let $(Pseudonym, m, AgY, YA)$ be a record of the product information m. The system adds it into the table $Table$.

Sign. When receiving a product, each S_{k_i} computes $Sig_i = H_1(C_1 \parallel C_2 \parallel m)^{h_i \cdot x_{S_{k_i}}}$. S_{k_i} sends Sig_i to $S_{k_{i+1}}$ for $i = 1, 2, 3, \cdots, d-2$. Finally, S_{k_d} computes Sig_d and $\sigma = \prod_{i=1}^{d} Sig_i$. Subsequently, S_{k_d} adds it into the record of m in the table $Table$.

User-Ownership-Verify. To prove the ownership of the product to the last logistics station S_{k_d}. U and S_{k_d} work as follows.

1) U selects $v_1 \xleftarrow{R} Z_q, v_2 \xleftarrow{R} Z_q$ and computes $V_1 = g_2^{v_1}, V_2 = Y_t^{v_1} \cdot g_2^{v_2}$.

2) U sends C_1, C_2, V_1, V_2 to S_{k_d}. S_{k_d} selects $c \xleftarrow{R} Z_q$, and returns it to U.

3) U computes $r_1 = v_1 - c \cdot k$, and $r_2 = v_2 - c \cdot x_u$, and returns (r_1, r_2) to S_{k_d}.

4) S_{k_d} verifies $V_1 \overset{?}{=} g_2^{r_1} \cdot C_1^c$, and $V_2 \overset{?}{=} Y_t^{r_1} \cdot g_2^{r_2} \cdot C_2^c$. If these equations hold, it outputs 1 to show that U is the owner of the product; otherwise, it outputs 0 to show that U is not the owner of the product.

Verify. U verifies $e(\sigma, g_2^{-1}) \cdot e(H_1(C_1 \parallel C_2 \parallel m), YA) \overset{?}{=} 1_{G_t}$. If the equation holds, it outputs 1 to show that the delivery process is correct; otherwise, it outputs 0 to show that there are some errors in the delivery.

Trace. In the case that the identity of U who selected the product m needs to be revealed, T searches on $Table$, and finds the record $(\sigma, Pseudonym, YA, m)$ firstly. Then, T verifies $e(\sigma, g_2^{-1}) \cdot e(H_1(C_1 \parallel C_2 \parallel m), YA) \overset{?}{=} 1_{G_t}$. If it is not, T quits the system immediately; otherwise, T computes $Y_u = C_2/C_1^{x_t}$, and confirms the identity of user.

4   Security Analysis

In this section, the security of our scheme is formally proven.

Theorem 1. *Our privacy-preserving logistics information system with traceability (PPLIST) is $(\varrho, \epsilon(l))-$ unforgeable if and only if the $(\epsilon(l)', T)-$ computational Diffie-Hellman (CDH) assumption holds on the bilinear group $(e, q, G_1, G_2, G_\iota)$ and H_1, H_2 are two random oracles and H_3 is a cryptographic hash function, where ϱ is the number of signature queries made by the forger $\mathcal{F}$, and $\epsilon(l)' \geq \frac{1}{q} \cdot \frac{1}{q-1} \cdot \frac{1}{\varrho} \cdot \epsilon(l)$.*

Proof. The proof of this theorem is referred to the full version of this paper [13].

Theorem 2. *Our privacy-preserving logistics information system with traceability (PPLIST) is $(\varrho, \epsilon(l))-$ traceable if the computational Diffie-Hellman (CDH) assumption holds on the bilinear group $(e, q, G_1, G_2, G_\iota)$ with the advantage at most $\epsilon_1(l)$, the discrete logarithm (DL) assumption holds on the group G_2 with the advantage at most $\epsilon_2(l)$, and H_1, H_2, H_3 are random oracles, where ϱ is the number of signature queries made by the forger $\mathcal{F}$, and $\epsilon(l) = max\{\frac{1}{2} \cdot \frac{1}{q} \cdot \frac{1}{\varrho} \cdot \epsilon_1(l), \frac{1}{2} \cdot \epsilon_2(l)\}$.*

Proof. The proof of this theorem is referred to the full version of this paper [13].

5   Experiment and Evaluation

In this section, we introduce the implementation and evaluation of our PPLIST scheme.

5.1   Runtime Environment

The performance of our PPLIST scheme is measured on a Lenovo Legion Y7000P 2018 laptop with an Intel Core i7-8750H CPU, 500 GB SSD and 8 GB RAM. The scheme is implemented in Microsoft Windows 10 System using E-clipse Integrated environment, Java language and JPBC library [3].

In our implementation, we apply the Type F curve. For the hash functions $H_1 : \{0,1\}^* \to G_1$, $H_2 : \{0,1\}^* \to Z_q$ and $H_3 : \{0,1\}^* \to Z_q$ required by our scheme, we used $SHA-256$ and the "newElementfromHash()" method in the JPBC library.

Our scheme is implemeted in the following three cases: 1) $n = 20, d = 10$; 2) $n = 100, d = 50$; 3) $n = 200, d = 100$. The experimental results are shown in Table 3.

Table 3. Times (ms)

Phase	n = 20, d = 10	n = 100, d = 50	n = 200, d = 100
Setup	522	519	506
Station-Key-Generation	137	578	1031
User-Key-Generation	7	5	4
Trace-Key-Generation	8	5	3
User-Pseudonym	27	12	12
Public-Key-Aggregation	217	745	1400
Sign	122	486	953
User-Ownership-Verify	112	106	97
Verify	176	144	141
Trace	186	144	141

5.2 Timing

The setup phase is a process run by the system. It takes 522 ms, 519 ms and 506 ms to setup the system in case 1, case 2 and case 3, respectively. According to the data, it can be observed that the running time of the three cases is roughly the same in the setup phase.

The key pair generation phase is run by logistics stations, user and trace party. It takes 137 ms, 578 ms and 1031 ms to generate the key pair of logistics stations in case 1, case 2 and case 3, respectively. In the user key pair generation phase, it takes 7 ms, 5 ms and 4 ms in case 1, case 2 and case 3, respectively. For the trace party to generate key pair, it takes 8 ms, 5 ms and 3 ms in case 1, case 2 and case 3, respectively.

The pseudonym generation phase is run by the user. It takes 27 ms, 12 ms and 12 ms in case 1, case 2 and case 3, respectively. Observing the experimental data, it is not difficult to find that the running time of the public key aggregation phase is proportional in the number of logistics stations. It takes 259 ms, 745 ms and 1400 ms to aggregate public keys in the three cases, respectively. The signature phase is run by logistics stations. The times to generate a multi-signature in case 1, case 2 and case 3 are 122 ms, 486 ms and 953 ms, respectively.

In the user ownership verification phase, a user proves the ownership by interacting with the last logistics station. It takes 112 ms, 106 ms and 97 ms in case 1, case 2 and case 3, respectively. In the signature validation phase, it takes 186 ms, 144 ms and 141 ms to verify a multi-signature in case 1, case 2 and case 3, respectively. To trace a user, it takes 176 ms, 144 ms and 141 ms in case 1, case 2 and case 3, respectively. We implement our scheme in three different cases. The experiment results show the efficiency of our scheme.

6   Conclusions

In this paper, to protect users' privacy in LIS, a PPLIST was proposed. In our scheme, users anonymously use logistics services. Furthermore, a trace party can de-anonymize users to prevent illegal logistics. Additionally, the whole logistics process can be recorded and is unforgeable. We formalized the definition and security model of our scheme, and presented a formal construction. We formally proved the security of our scheme and implemented it.

In our scheme, a buyer can prove the ownership of a product by proving the knowledge included in the pseudonyms. Our future work is to improve the flexibility of this work to enable an owner to designate a proxy to prove the ownership of products on behalf of him.

Acknowledgment. This work was partially supported by the National Natural Science Foundation of China (Grant No. 61972190, 62072104, 61972095) and the National key research and development program of China (Grant No. 2020YFE0200600). This work was also partially supported by the Postgraduate Research & Practice Innovation Program of Jiangsu Province (Grand No. KYCX20_1322) and the Natural Science Foundation of the Fujian Province, China (Grant No. 2020J01159).

References

1. Amazon. https://www.amazon.cn
2. Taobao. https://www.taobao.com
3. Angelo, D.C., Vincenzo, I.: JPBC: Java pairing based cryptography. In: ISCC 2011, pp. 850–855. IEEE (2011)
4. Asaar, M.R., Salmasizadeh, M., Susil, W.: An identity-based multi-proxy multi-signature scheme without bilinear pairings and its variants. Comput. J. **58**(4), 1021–1039 (2015)
5. Bagherzandi, A., Cheon, J.H., Jarecki, S.: Multisignatures secure under the discrete logarithm assumption and a generalized forking lemma. In: CCS 2008, pp. 449–458. ACM (2008)
6. Bardi, E.J., Raghunathan, T.S., Bagchi, P.K.: Logistics information systems: the strategic role of top management. J. Bus. Logist. **15**(1), 71–85 (1994)
7. Bellare, M., Neven, G.: Multi-signatures in the plain public-key model and a general forking lemma. In: CCS 2006, pp. 390–399. ACM (2006)
8. Boldyreva, A.: Threshold signatures, multisignatures and blind signatures based on the gap-Diffie-Hellman-group signature scheme. In: Desmedt, Y.G. (ed.) PKC 2003. LNCS, vol. 2567, pp. 31–46. Springer, Heidelberg (2003). https://doi.org/10.1007/3-540-36288-6_3
9. Boneh, D., Drijvers, M., Neven, G.: Compact multi-signatures for smaller blockchains. In: Peyrin, T., Galbraith, S. (eds.) ASIACRYPT 2018. LNCS, vol. 11273, pp. 435–464. Springer, Cham (2018). https://doi.org/10.1007/978-3-030-03329-3_15
10. Chaum, D.: Security without identification: transaction systems to make big brother obsolete. Commun. ACM **28**(10), 1030–1044 (1985)

11. Chaum, D., Evertse, J.-H.: A secure and privacy-protecting protocol for transmitting personal information between organizations. In: Odlyzko, A.M. (ed.) CRYPTO 1986. LNCS, vol. 263, pp. 118–167. Springer, Heidelberg (1987). https://doi.org/10.1007/3-540-47721-7_10

12. Chen, L.: Access with pseudonyms. In: Dawson, E., Golić, J. (eds.) CPA 1995. LNCS, vol. 1029, pp. 232–243. Springer, Heidelberg (1996). https://doi.org/10.1007/BFb0032362

13. Chen, Q., Han, J., Li, J., Chen, L., Li, S.: A privacy-preserving logistics information system with traceability (2021). https://arxiv.org/abs/2109.05216

14. Chrisos, M.: Pseudonymization Techniques: How to Protect Your Data (2019). https://www.techfunnel.com/information-technology/pseudonymization-techniques-how-to-protect-your-data/

15. Christin, N., Yanagihara, S.S., Kamataki, K.: Dissecting one click frauds. In: CCS 2010, pp. 15–26. ACM (2010)

16. Closs, D.J., Xu, K.: Logistics information technology practice in manufacturing and merchandising firms - an international benchmarking study versus world class logistics firms. Int. J. Phys. Distrib. Logist. Manag. **30**(10), 869–886 (2000)

17. Correll, N., et al.: Analysis and observations from the first Amazon picking challenge. IEEE Trans. Autom. Sci. Eng. **15**(1), 172–188 (2018)

18. Gao, Q., Zhang, J., Ma, J., Yang, C., Guo, J., Miao, Y.: LIP-PA: a logistics information privacy protection scheme with position and attribute-based access control on mobile devices. Wirel. Commun. Mob. Comput. **2018**(1), 1–14 (2018)

19. Gordon, D.M.: Discrete logarithms in GF(P) using the number field sieve. SIAM J. Discrete Math. **6**(1), 124–138 (1993)

20. Grinshpoun, T., Grubshtein, A., Zivan, R., Netzer, A., Meisels, A.: Asymmetric distributed constraint optimization problems. J. Artif. Intell. Res. **47**(1), 613–647 (2013)

21. Han, J., Chen, L., Schneider, S., Treharne, H., Wesemeyer, S., Wilson, N.: Anonymous single sign-on with proxy re-verification. IEEE Trans. Inf. Forensics Secur. **15**(1), 223–236 (2020)

22. Hanaoka, G., Kurosawa, K.: Efficient chosen ciphertext secure public key encryption under the computational Diffie-Hellman assumption. In: Pieprzyk, J. (ed.) ASIACRYPT 2008. LNCS, vol. 5350, pp. 308–325. Springer, Heidelberg (2008). https://doi.org/10.1007/978-3-540-89255-7_19

23. Horster, P., Michels, M., Petersen, H.: Meta-multisignature schemes based on the discrete logarithm problem. In: Information Security—the Next Decade. IAICT, pp. 128–142. Springer, Boston (1995). https://doi.org/10.1007/978-0-387-34873-5_11

24. Itakura, K., Nakamura, K.: A public-key cryptosystem suitable for digital multisignatures. NEC Res. Dev. **71**(1), 1–8 (1983)

25. Kang, J., Yu, R., Huang, X., Zhang, Y.: Privacy-preserved pseudonym scheme for fog computing supported internet of vehicles. IEEE Trans. Intell. Transp. Syst. **19**(8), 2627–2637 (2018)

26. Kim, D.J., Song, Y.I., Braynov, S.B., Rao, H.R.: A multidimensional trust formation model in B-to-C e-commerce: a conceptual framework and content analyses of academia/practitioner perspectives. Decis. Support Syst. **40**(2), 143–165 (2005)

27. Korth, B., Schwede, C., Zajac, M.: Simulation-ready digital twin for realtime management of logistics systems. In: IEEE Big Data 2018, pp. 4194–4201. IEEE (2018)

28. Lai, K., Ngai, E., Cheng, T.: Information technology adoption in Hong Kong's logistics industry. Transp. J. **44**(4), 1–9 (2005)

29. Liu, F., Shu, P., Lui, J.C.S.: *AppATP*: an energy conserving adaptive mobile-cloud transmission protocol. IEEE Trans. Comput. **64**(11), 3051–3063 (2015)
30. Liu, S., Wang, J.: A security-enhanced express delivery system based on NFC. In: ICSICT 2016, pp. 1534–1536. IEEE (2016)
31. Lysyanskaya, A., Micali, S., Reyzin, L., Shacham, H.: Sequential aggregate signatures from trapdoor permutations. In: Cachin, C., Camenisch, J.L. (eds.) EUROCRYPT 2004. LNCS, vol. 3027, pp. 74–90. Springer, Heidelberg (2004). https://doi.org/10.1007/978-3-540-24676-3_5
32. Lysyanskaya, A., Rivest, R.L., Sahai, A., Wolf, S.: Pseudonym systems. In: Heys, H., Adams, C. (eds.) SAC 1999. LNCS, vol. 1758, pp. 184–199. Springer, Heidelberg (2000). https://doi.org/10.1007/3-540-46513-8_14
33. Léauté, T., Faltings, B.: Coordinating logistics operations with privacy guarantees. In: IJCAI 2011, pp. 2482–2487. Morgan Kaufmann (2011)
34. Marko, A., Marjan, M., Vlada, S.: Logistics information system. Vojnotehnicki glasnik **58**(1), 33–61 (2010)
35. Micali, S., Ohta, K., Reyzin, L.: Accountable-subgroup multisignatures: extended abstract. In: CCS 2001, pp. 245–254. ACM (2001)
36. Narayanan, A., Shmatikov, V.: De-anonymizing social networks. In: S&P 2009, pp. 173–187. IEEE (2009)
37. Neven, G.: Efficient sequential aggregate signed data. IEEE Trans. Inf. Theory **57**(3), 1803–1815 (2011)
38. Ngai, E., Lai, K.H., Cheng, T.: Logistics information systems: the Hong Kong experience. Int. J. Prod. Econ. **113**(1), 223–234 (2008)
39. Ohta, K., Okamoto, T.: A digital multisignature scheme based on the Fiat-Shamir scheme. In: Imai, H., Rivest, R.L., Matsumoto, T. (eds.) ASIACRYPT 1991. LNCS, vol. 739, pp. 139–148. Springer, Heidelberg (1993). https://doi.org/10.1007/3-540-57332-1_11
40. Ohta, K., Okamoto, T.: Multisignature schemes secure against active insider attacks. IEICE Trans. Fundam. Electron. Commun. Comput. Sci. **E82-A**(1), 21–31 (1999)
41. Okamoto, T.: A digital multisignature scheme using bijective public-key cryptosystems. ACM Trans. Comput. Syst. **6**(4), 432–441 (1988)
42. Qi, H., Chenjie, D., Yingbiao, Y., Lei, L.: A new express management system based on encrypted QR code. In: ICICTA 2015, pp. 53–56. IEEE (2015)
43. Qian, J., Li, X., Zhang, C., Chen, L.: De-anonymizing social networks and inferring private attributes using knowledge graphs. In: INFOCOM 2016, pp. 1–9. IEEE (2016)
44. Qu, Z., He, P., Hou, L.: Studies on internet real-name system and network action surveillance system. In: EDT 2010, pp. 469–472. IEEE (2010)
45. Stallings, W.: Handling of personal information and deidentified, aggregated, and pseudonymized information under the California consumer privacy act. IEEE Secur. Priv. **18**(1), 61–64 (2020)
46. Tarjan, L., Senk, I., Tegeltija, S., Stankovski, S., Ostojic, G.: A readability analysis for QR code application in a traceability system. Comput. Electron. Agric. **109**(4), 1–11 (2014)
47. Tiwari, N., Padhye, S., He, D.: Efficient ID-based multiproxy multisignature without bilinear maps in ROM. Ann. Telecommun. **68**(3–4), 231–237 (2013). https://doi.org/10.1007/s12243-012-0315-x
48. Wei, Q., Li, X.: Express information protection application based on K-anonymity. Appl. Res. Comput. **31**(2), 555–567 (2014)

49. Wei, Q., Wang, C., Li, X.: Express information privacy protection application based on RSA. Appl. Electron. Tech. **40**(7), 58–60 (2014)
50. Xu, F.J., Tan, C.J., Tong, F.C.: Auto-ID enabled tracking and tracing data sharing over dynamic B2B and B2G relationships. In: RFID-TA 2011, pp. 394–401. IEEE (2011)
51. Ye, L., Wang, Y., Chen, J.: Research on the intelligent warehouse management system based on near field communication (NFC) technology. Int. J. Adv. Pervasive Ubiquit. Comput. **8**(2), 38–55 (2016)
52. Yong, L., Jing, P., Ma, B., Bo, Y.: The analysis of logistic bottleneck of Taobao.com in C2C model and its countermeasures. In: ICEE 2010, pp. 5537–5539. IEEE (2010)

Post-quantum Key Escrow for Supervised Secret Data Sharing on Consortium Blockchain

Xiaowen Cai[1], Wenjing Cheng[1], Minghui Zhang[1], Chen Qian[1], Zhengwei Ren[2], Shiwei Xu[1(✉)], and Jianying Zhou[3]

[1] College of Informatics, Huazhong Agricultural University, Wuhan, People's Republic of China
xushiwei@mail.hzau.edu.cn
[2] School of Computer Science and Technology, Wuhan University of Science and Technology, Wuhan, People's Republic of China
[3] Singapore University of Technology and Design, Singapore, Singapore

Abstract. Consortium blockchain has been widely used in different management scenarios (i.e., digital finance), where normal members want to keep their on-chain data private while supervision peers want to reveal the on-chain private data under certain circumstances like financial regulation and judicial forensics, and key escrow is an idea to solve the problem. Since current key escrow schemes heavily rely on traditional asymmetric encryption and decryption algorithms that are vulnerable to attacks from quantum computers, we design and implement the first post-quantum (PQ) key escrow system for consortium blockchains (i.e., PQ-KES4Chain), which is integrated with all the PQ public-key encryption/KEM candidate algorithms in the current round of NIST call for national standard. Furthermore, we provide chaincodes, related APIs together with client codes for further development. And we perform a detailed security analysis on the system design and a full evaluation on the performance of PQ-KES4Chain including the time of chaincodes execution and the on-chain storage space. We further discuss the implications of our findings, which could be helpful for the developers of PQ KEM algorithms and applications.

Keywords: Consortium blockchain · Key escrow protocol · Supervised secret data sharing · Post quantum cryptography

1 Introduction

Consortium blockchain has been widely used by groups of organizations or corporations, which want to enjoy the advantages in sharing data and information offered by distributed ledgers. As a kind of distributed ledger, consortium blockchain has notable advantages like better data integrity, greater transparency, enhanced traceability and so on. However, the confidentiality of the on-chain data can be easily violated because of the transparency and traceability offered by consortium blockchain.

To preserve the confidentiality of the on-chain data, the most straightforward method is making use of cryptographic encryption and decryption. Due to encryption speed, the

© Springer Nature Switzerland AG 2022
W. Meng and M. Conti (Eds.): CSS 2021, LNCS 13172, pp. 164–181, 2022.
https://doi.org/10.1007/978-3-030-94029-4_12

data uploader normally asymmetrically (resp. symmetrically) encrypts small amount of data like session keys (resp. big amount of data), and the data downloader decrypts the encrypted data according to the corresponding encryption algorithm. Nevertheless, in that case, the encrypted data become an on-chain enclave that is outside of the consortium administrator's supervision. The enclave may violate the transparency and traceability of blockchain, which is badly needed in the scenarios like financial regulation and judicial forensics.

Some consortium blockchains (e.g., Hyperledger Fabric [1] and Quorum [2]) also provide private channel (or peer networks) for a group of admitted peers, in order to prevent unadmitted participants from accessing the related data protected by the channel. Whereas, if the data are exchanged between peers inside and outside the private channel, then there is no guarantee of the data confidentiality.

Key Escrow [3] is an idea that allows authorized entities (e.g., government officials) to decrypt encrypted data under certain conditions (e.g., after getting a court permit). Consequently, key escrow provides a solution, which considers both data confidentiality and supervised data disclosure, for the consortium blockchain.

In a typical key escrow protocol, it heavily relies on asymmetric encryption and decryption to protect and recover the session key, which is used to further protect the shared secret data. However, since the quantum computers are coming and the related Shor's algorithm [4] is devastating attack to traditional asymmetric cryptography (e.g., RSA, ECDSA, ECIES), PQ asymmetric cryptography is actively being developed as a substitution for traditional asymmetric cryptography. Among plenty of research projects and standardization initiatives for PQ cryptography, it is worth noting the National Institute of Standards and Technology (NIST) call for proposals of PQ public-key cryptosystems including public-key encryption/Key Encapsulation Mechanism (KEM) and digital signature algorithms, which is currently in its third round [5] and expected to deliver the first national standard drafts between 2021 and 2022. For future consideration, it is meaningful to evaluate the existing PQ public-key encryption/KEM algorithms in real scenarios such as key escrow system based on the consortium blockchain.

1.1 Research Contribution

In this paper, we propose the first PQ key escrow system for consortium blockchain, which guarantees both data confidentiality and supervised data disclosure. The contributions of our work are summarized as follows.

- We design the key escrow system for consortium blockchain mainly based on smart contracts without relying on the protection of any cryptographic chips. Instead, the underlying consortium blockchain can provide guarantee for the integrity of both escrow-related data and chaincode execution. Furthermore, we integrate the system with all the PQ public-key encryption/KEM algorithms in the current round of NIST call for national standard.
- We implement our proposed system on top of Hyperledger Fabric, and provide chaincodes, related APIs together with client codes, which allow application developers to further create their PQ supervised secret data sharing applications.

- We conduct comprehensive security analysis on PQ-KES4Chain and emphasize how to reach required PQ security level when using the PQ KEM algorithms in real application scenarios (e.g., key escrow system).
- We perform a full evaluation on the performance of PQ-KES4Chain including the consumed time of chaincodes execution and the needed on-chain storage space, and then discuss the implications of our findings, which could be helpful for the PQ KEM application developers and the developers of PQ KEM algorithms.

To the best of our knowledge, our work is also the first to conduct a full evaluation on the performance and security of all the NIST candidate public-key encryption/KEM algorithms in blockchain-based scenarios (e.g., key escrow system).

1.2 Related Works

The first key escrow protocol is Escrow Encryption Standard (EES) [6] standardized by NIST, and is implemented on a hardware chip called Clipper to protect keys and codes stored in each peer involved in the protocol. Following the idea of EES, some works developed software-based key escrow protocols [7–9] by making use of asymmetric cryptography. However, these systems assume that fully trusted third parties provide key escrow services. To limit the power of such trusted escrow agents, researchers proposed Partial Key Escrow (PKE) [10, 11] and Encapsulated Key Escrow (EKE) [12], which both tried to computationally prohibit the trusted escrow agents from launching large-scale wiretapping by imposing a time delay between obtaining the escrowed information of a user and recovering the decryption secret key.

Some recent works have deployed key escrow to the blockchain system for supervised data disclosure and data separation. Goldfeder *et al.* [13] proposed escrow services for cryptocurrencies transactions using Bitcoin in untrusted environments, where a multi-signature based on the (t, n) threshold cryptosystem is utilized. Cha *et al.* [14] applied key escrow system to blockchain so that someone who requested information disclosure can check the information under the smart contract. In [15], a key escrow is employed for the prover, who acquires anonymous and trusted location proofing on the blockchain-based system, to achieve the separation of identity information and location information. Nevertheless, none of these key-escrow works considers threat from quantum computers.

Currently, the NIST call for proposals of PQ public-key cryptosystems is in its third round, there are six candidate digital signature algorithms and nine candidate public-key encryption/KEM algorithms left. Based on the candidate digital signature algorithms in the call, many researchers [16–19] have developed PQ blockchain systems. Whereas, on the other hand, only few work [20] considers the candidate public-key encryption/KEM algorithms, and lacks of a scheme for applying the candidate algorithms to blockchain systems, not even to mention a full evaluation on the security and performance of the algorithms running on blockchain.

1.3 Paper Organization

The rest of the paper is organized as follows. In Sect. 2, we introduce background information about consortium blockchains, key escrow and PQ cryptography. And then we present the design and implementation of our PQ key escrow system based on consortium blockchain in Sect. 3. After that we comprehensively analyze the security of the system design and implementation in Sect. 4 and evaluate the performance of our system in Sect. 5. Finally, we draw a conclusion in Sect. 6.

2 Preliminaries

In this section, we take an overview of consortium blockchains and Hyperledger Fabric. And then we introduce the idea of key escrow and the example protocol of key escrow. Finally, we review all the candidate public-key encryption/KEM algorithms in the third round of NIST call for proposals of PQ cryptography.

2.1 Consortium Blockchain and Hyperledger Fabric

Consortium blockchain operates under a set of organizations, and provides a way to secure the interactions among the organizations that have a common goal but do not fully trust each other, just like peers sharing secret data in/between organizations and the data supervision organization.

Among current existing consortium blockchains, Hyperledger Fabric is one of the most successful projects, and it supports executing distributed applications (i.e., chaincodes, aka smart contracts) on a consortium of organizations and peers in one channel, whose business logic is programmed as chaincodes written in standard, general-purpose programming languages (e.g., Go, Node.js, Java).

As shown in Fig. 1, Hyperledger Fabric introduces an *execute-order-validate* architecture, where *endorsement peer*, *commitment peer*, *orderer* and *client* are deployed, and peers can be grouped by the organizations they belong to. All the chaincodes should be instantiated on specific peers before the execution of Hyperledger Fabric. The work flow of Hyperledger Fabric is introduced as follows. Firstly, the client invokes the execution of chaincode by submitting a transaction proposal to the endorsement peers. Secondly, after executing chaincode and generating signature for the execution result, the endorsement peers send the endorsed proposals back to the client. Thirdly, the client then broadcasts the endorsed proposals to the orderers. Fourthly, the orderers establish a total order on all submitted transactions in one channel, batch these transactions into a new block, and distribute both to all the commitment peers in the channel. Finally, the commitment peers validate each transaction in the new block, and append the block to the distributed ledger.

During the execution of Hyperledger Fabric, peers can access on-chain data by *keys*, which are like member variables bound to instantiated chaincodes, and a *key* can be a single datum or a tuple of data. All modifications/updates to the data and instantiated chaincodes are recorded by Hyperledger Fabric and can be easily supervised.

Besides the on-chain data, peers and organizations may have secret data kept in their own off-chain private databases. When the secret data are shared online, peers and organizations upload the encrypted data to keep their data secret from other unexpected participants, which may violate the transparency and traceability of blockchain. Therefore, this drives the development of supervised secret data sharing based on key escrow, which balances the needs of on-chain data secrecy and transparency.

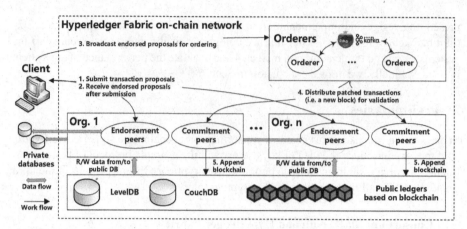

Fig. 1. The architecture, work flow and data flow of Hyperledger Fabric.

2.2 Key Escrow Systems and Example Protocol

The original idea of key escrow is proposed by Silvio Micali in his work about fair public-key cryptosystems [21]. In key escrow systems, exchanged secret data are encrypted under pre-negotiated session keys and every session key used in the communication between *users* is escrowed (e.g., asymmetrically encrypted under the escrow key). Normally, the escrowed session keys can be recovered (e.g., asymmetrically decrypted using the corresponding escrow key) by law enforcement agents named *escrow agents*, which will further decrypt the secret data using the recovered session key (in the case of financial regulation and judicial forensics). If one escrow agent is not fully trusted, the system should be configured as multiple escrow agents. Therefore, one key escrow system consists of two groups of entities (i.e., *users* sharing a session key used to protect the exchanged secret data and *escrow agents* recovering the session key).

We take the software-based key escrow protocol from [7] as an example, which involves two users and two escrow agents. In the example protocol, the two key escrow agents *KEA1* and *KEA2* separately have their own public/private key pairs denoted as (*KEA1Pub*, *KEA1Priv*) and (*KEA2Pub*, *KEA2Priv*), of which the two public keys are both integrated with the key escrow programs (i.e., P_A and P_B) separately executed by the two users A and B. The execution of the example key escrow protocol is introduced as follows.

1) Suppose user A wants to send a message M to user B. Firstly, A and B need to pre-negotiate a session key SK.
2) Then, A feeds M and SK to his/her key escrow program P_A, which will output the following information: (1) the symmetrically encrypted message $C = SEnc_{SK}(M)$ under the session key SK, (2) the catenation of two randomly-generated temporary session key SK_1 and SK_2 asymmetrically encrypted under the two key escrow agents' public key, whose exclusive-or result is the session key SK. And the catenation is denoted as $LEAF = AEnc_{KEA1Pub}(SK_1) \; || \; AEnc_{KEA2Pub}(SK_2)$ where $SK = SK_1 \oplus SK_2$.

3) Upon receiving the above messages from user A, user B will feed them together with the session key SK to his/her key escrow program P_B, which will decrypt the encrypted message C by using the session key SK to obtain the message M.

4) To wiretap the communication between user A and user B, one supervision department needs to present a court order together with the $LEAF$ information to the two key escrow agents $KEA1$ and $KEA2$, who will separately decrypt the corresponding components using their own private keys and give SK_1 and SK_2 back to the supervision department. Finally, the supervision department uses $SK_1 \oplus SK_2$ to decrypt the encrypted message C and finally gets the message M.

As one may notice, the crucial point of implementing the software-based key escrow system is to ensure that the key escrow programs (e.g., P_A and P_B) execute correctly and not be modified by malicious users, who want to bypass or corrupt the escrow system. Fortunately, consortium blockchains can provide guarantee for the integrity and traceability of the data, identity, and codes execution to solve the above problem.

2.3 Post-quantum Public-Key Encryption/KEM Algorithms in the NIST Call

In the third (current) round of NIST call for PQ public-key encryption/KEM algorithms, there are four finalists (i.e., Classic McEliece [22], CRYSTALS-KYBER [23], NTRU [24], SABER [25]) and five alternate candidates (i.e., BIKE [26], NTRU Prime [27], FrodoKEM [28], SIKE [29], HQC [30]). The finalists will continue to be reviewed for consideration for standardization at the conclusion of the third round, while alternate candidates are just potentially standardized (most likely will not occur at the end of the third round) but may be reconsidered for various reasons (e.g., better performance, higher security level, broader range of hardness assumptions) in a fourth round of evaluation held by NIST.

To eliminate the complexity of the padding scheme and the proofs needed to show the padding is secure, NIST currently only announces the KEM mode (rather than the encryption/decryption mode) of the PQ public-key encryption/KEM algorithms. And one key encapsulation mechanism KEM consists of following 3 steps.

- The key generation step $KEM.KeyGen()$, that outputs a public/private key pair $(pubKey, privKey)$.
- The encapsulation step $KEM.Encap(pubKey)$ that takes a public key $pubkey$ as input, and outputs a shared secret/ciphertext pair (SS, CT).
- The decapsulation step $KEM.Decap(privKey, CT)$, that takes the corresponding private key $privKey$ and the ciphertext CT as input, and outputs the shared secret SS.

To sum up, in a KEM-based protocol, Alice generates a shared secret SS and a ciphertext CT that encapsulates SS, and Bob decapsulates SS from CT. The shared secret SS is further used as a symmetric key to encrypt/decrypt the exchanged secret data between Alice and Bob.

As shown in Table 1, we summarize the details of all the finalists and alternate candidates in NIST third-round call for PQ public-key encryption/KEM algorithms.

Table 1. Details of PQ KEM algorithms in NIST third-round call

Algorithms	Claimed NIST security level	Public key size (bytes)	Private key size (bytes)	Ciphertext size (bytes)	Shared secret size (bytes)
Classic McEliece	1, 3, 5	261120~1357824	6452~14080	128~240	32
CRYSTALS-KYBER	1, 3, 5	800~1568	1632~3168	768~1568	32
NTRU	1, 3, 5	699~1138	935~1450	699~1138	32
SABER	1, 3, 5	672~1312	1568~3040	736~1472	32
BIKE	1, 3	2542~6206	3110~13236	2542~6206	32
FrodoKEM	1, 3, 5	9616~21520	19888~43088	9720~21632	16, 24, 32
HQC	1, 3, 5	2249~7425	2289~7285	4481~14469	64
NTRU Prime	2, 3, 4	897~1322	1125~1999	897~1184	32
SIKE	1, 2, 3, 5	197~564	350~644	236~596	16, 24, 32

Due to space limit, we refer the reader to [5] for more details. By using different key sizes, all the algorithms can achieve different NIST security levels (i.e., bits-of-security level), which is defined as the effort required by a classical computer to perform a brute-force attack on a given-length cryptographic key. Normally NIST security levels 1–5 approximately imply 128/160/192/224/256-bits-of security levels. Compared to the shared secret size, all the algorithms have relatively big public key, private key and ciphertext sizes, which indicate that a PQ key escrow system would be storage-consuming.

3   System Design and Implementation

3.1   System Architecture and Execution Flow

Figure 1 shows the architecture of PQ-KES4Chain, where we design the key escrow system for consortium blockchain mainly based on smart contracts without relying on the protection of any cryptographic chips. Instead, the underlying consortium blockchain can provide guarantee for the integrity of the escrow-related data, while on the other hand, developers should pre-record and check the version numbers of all the expected chaincodes in the client codes to prevent malicious peers from modifying or counterfeiting the installed chaincodes.

Peers in PQ-KES4Chain are divided into three kinds of groups namely data sharing group (organization), escrow agent group (organization) and supervisor group (organization). The first organization includes sender/receiver peers that are willing to escrow their session key to escrow agents. Whereas the escrow agent organization and the supervisor organization separately consist of one escrow agent peer and one supervisor peer, where session key is escrowed to escrow agent peer and the supervisor peer may obtain the recovered session key from the escrow agent peer. In case of untrusted escrow agent, PQ-KES4Chain is configured as two escrow agents (peers) in two escrow agent organizations. Furthermore, to prevent escrow agent peers from reading each other's on-chain data prepared for the supervisor peer, we separate the two escrow agent peers/organizations

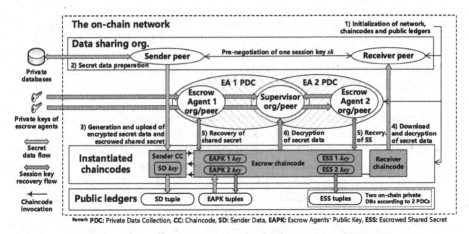

Fig. 2. The architecture, work flow and data flow of PQ-KES4Chain.

by using two on-chain private databases (i.e., Private Data Collection [31], PDC). We will comprehensively analyze the security design and implementation of PQ-KES4Chain in Sect. 4.

Since NIST only offers the KEM mode of all the related PQ algorithms, we need to design the corresponding execution flow of the key escrow system. As shown in Fig. 1, the execution flow of PQ-KES4Chain consists of six phases: 1) *initialization*, 2) *secret data preparation*, 3) *generation and upload of encrypted secret data and escrowed shared secret* (by sender peer), 4) *download and decryption of secret data* (by receiver peer), 5) *recovery of shared secret* (by key escrow agent peers) and 6) *decryption of secret data* (by supervisor peer).

1) *Initialization*: We deploy all the organizations and peers, and then instantiate three chaincodes (i.e., `Receiver` chaincode, `Sender` chaincode and `Escrow` chaincode) on specific peers, where the first two chaincodes are installed on every peer in data sharing organization while the last two chaincodes are installed on every peer in escrow agent organizations and supervisor organization. Firstly, to protect the exchanged secret data, PQ-KES4Chain requires sender and receiver peers to pre-negotiate one session key *SK* (by an either off-chain method or on-chain session based on the PQ KEM algorithms). Moreover, it is needed to initially invoke the `Escrow` chaincode to generate one PQ public/private key pair for each escrow agent (denoted as $pubKey_1/privKey_1$ and $pubKey_2/privKey_2$), which are used to escrow and recover the shared secret between sender peer and escrow peer. Furthermore, since the on-chain data tuples are accessed via *keys*, one send-receive-supervise session needs an offline agreement in advance on the keys of sender data and escrow agents' public keys (denoted as SD *key* and EAPK *keys*), which are like member variables separately bound to the `Sender` chaincode and `Escrow` chaincode. The SD *key* points to the encrypted secret data and the escrowed shared secret generated by the sender peer.

2) *Secret data preparation*: After the initialization step, the sender peer reads the secret data M from its off-chain private database and prepare for the key escrow process.

3) *Generation and upload of encrypted secret data and escrowed shared secret*: Once the secret data are ready, the sender peer invokes the `Sender` chaincode and performs the following actions.

- Firstly, the `Sender` chaincode symmetrically encrypts the secret data M using the pre-negotiated session key SK (denoted as $C = SEnc_{SK}(M)$).
- After that, because NIST only offers the KEM mode of all the related PQ algorithms, sender peer performs the encapsulation operation (by using the two escrow agents' public keys $pubKey_1$ and $pubKey_2$ under the EAPK *keys* bound to the `Escrow` chaincode), which generates two escrowed shared secrets (denoted as ESS_1 and ESS_2) and two ciphertexts (denoted as CT_1 and CT_2).
- Then, the sender peer symmetrically encrypts the secret data M using the exclusive-or result of ESS_1 and ESS_2 as the key (i.e., $C' = SEnc_{ESS}(M)$ where $ESS = ESS_1 \oplus ESS_2$).
- Finally, the sender peer uploads the catenation of C, C', CT_1 and CT_2 to the sender data tuple under SD *key* bound to the `Sender` chaincode. Due to the characteristics of Hyperledger Fabric, the receiver and escrow agent peer can only retrieve the sender data by invoking the `Sender` chaincode.

4) *Download and decryption of secret data*: After the encrypted secret data C being generated and uploaded, the receiver peer invokes the `Receiver` chaincode to decrypt the secret data by using the pre-negotiated session key SK and finally obtains the exchanged secret data M (i.e., $M = SDec_{SK}(C)$).

5) *Recovery of session key/shared secret*: When disclosure of the exchanged secret data is needed, the two escrow peers invoke the `Escrow` chaincode to separately recover the escrowed shared secrets ESS_1 and ESS_2 using their own private keys $privKey_1$ and $privKey_2$ together with the ciphertexts CT_1. and CT_2 stored under the SD *key*. The decapsulation of escrowed shared secrets are denoted as $ESS_1 = Decap(privKey_1, CT_1)$ and $ESS_2 = Decap(privKey_2, CT_2)$. The ESS_1 and ESS_2 are then written to the ESS *keys* bound to the `Escrow` chaincode.

6) *Decryption of secret data*: Once the escrowed shared secrets ESS_1 and ESS_2 are ready, the supervisor peer invokes the `Escrow` chaincode to recover ESS by calculating the exclusive-or result of ESS_1 and ESS_2, reads the symmetrically encrypted message $C' = SEnc_{ESS}(M)$ from SD *key* bound to `Sender` chaincode, and finally decrypts C' to get the exchanged secret data M.

3.2 System Implementation

The main implementation of PQ-KES4Chain consists of three chaincodes, namely `Sender`, `Receiver`, and `Escrow`. As shown in Table 2, the `Sender` chaincode provides API to generate/upload encrypted secret data and escrowed shared secret (i.e., sender data) and API to retrieve the sender data for other chaincodes. The `Receiver` chaincode only provides API to decrypt the encrypted secret data, whereas the `Escrow` chaincode can be invoked to generate the escrow agent's public/private key pair, get the related public key, use the private key to decapsulate/recover the shared secret, and utilize the shared secret to decrypt the secret data.

To further help developers to create their PQ supervised secret data sharing applications, we also provide client codes, which are also summarized in Table 2, showing how to invoke the chaincodes.

Table 2. The chaincodes and client codes of PQ-KES4Chain.

Chaincodes	APIs	Descriptions
Sender	Gen_Sender _Data	Retrieve the exchanged secret data from private database, generate and upload sender data under Sender Data (SD) *key*
	Get_Sender _Data	Read sender data under the SD *key* from public ledger
Receiver	Dec_Sec _Data	Decrypt encrypted secret data read from SD *key* bound to the Sender chaincode
Escrow	Gen_EA_ KeyPair	Generate one PQ public/ private key pair for escrow agent, upload the public key to EAPK *key*, and store the private key off-chain
	Get_EA_ PubKey	Read the public key of escrow agent from EAPK *key*
	Decap_ Shared_Sec	Decapsulate (recover) the shared secret under the SD *key* used as the encryption key of the exchanged secret data
	Dec_Sec_ Data	Decrypt the secret data encrypted by the shared secret

Client codes	Descriptions
Sender	Invoke Sender chaincode to generate and upload of encrypted secret data and escrowed shared secret
Receiver	Invoke Receiver chaincode to download and decrypt the secret data
Escrow1_0 Escrow2_0	Invoke Escrow chaincode to generate the PQ public/private key pair for escrow agent 1/2
Escrow1_1 Escrow2_1	Invoke Escrow chaincode to decapsulate (recover) the shared secret escrowed to escrow agent 1/2
Supervis- or	Invoke Escrow chaincode to decrypt the secret data using the decapsulated shared secret
CAUtil AppUtil	Invoked by the other client codes to get the Fabric CA certificates and setup information of our system

As shown in Table 2, there are nine client codes available, namely Sender, Receiver, Escrow1_0, Escrow2_0, Escrow1_1, Escrow2_1, Supervisor, CAUtil and AppUtil, of which the former two invoke the corresponding chaincode APIs and the following five client codes invoke Escrow chaincode to separately generate PQ key pair for escrow agent 1/2, decapsulate the shared secret escrowed to escrow agent 1/2 and decrypt the secret data using the decapsulated shared secret. The last two client codes are invoked by the other client codes to get the Hyperledger Fabric CA certificates and setup information of our system.

A detailed analysis on security design and implementation of the chaincodes and client codes is presented in Sect. 4, and all the codes are available on [32].

4   Security Analysis

In this section, we introduce the threat model of PQ-KES4Chain, and then present how we design/implement PQ-KES4Chain to protect on-chain data and chaincodes from attacks in the threat model. Finally, we analyze how to use PQ KEM algorithms together with other cryptographic algorithms in real application scenarios such as our key escrow system in order to achieve enough security level under the attacks from quantum computers.

4.1   Threat Model

All the peers in PQ-KES4Chain may try to perform attacks to bypass or corrupt the key escrow system. Possible attacks from malicious peers are listed as follows.

- **Malicious sender/receiver peer** may want to modify or substitute the sender data under the **SD** *key* belonging to another sender-receiver session, and may want to read the escrowed shared secret under the **ESS** *key* that should not be exposed outside the escrow agent organization.
- **Malicious escrow peer** may try to recover or modify the escrowed shared secret under the **ESS** *key* that is escrowed to another escrow peer.
- **Malicious supervisor peer** may want to recover the escrowed shared secret under the **ESS** *key* (to get the session key and further decrypt the secret data) without the help of escrow agents.

Furthermore, **all malicious peers** may want to generate or substitute the escrow agent public key under the **EAPK** *key* in order to mislead sender/receiver peer using the rogue public key, and may also want to modify/update one **chaincode** to bypass the security check in the original chaincode.

4.2   Security Design

To save PQ-KES4Chain from the attacks that malicious peers may perform, we develop and deploy security mechanisms to protect the integrity and secrecy of the on-chain data and chaincodes.

- **Integrity of sender data under SD *key*.** For sake of simplicity, we do not check the integrity of the sender data. However, one developer can easily perform this check by calculating one Message Authentication Code (MAC) based on the session key SK (and timestamp), attaching the MAC with sender data and verifying the MAC. And the MAC can also be used to authenticate the identity of the sender.
- **Integrity of escrow agent's public key under the EAPK *key*.** We perform access control in the Escrow chaincode, to which the EAPK *key* is bound, by using the *getCreator*() API (in the *shim* package) in order to make sure that only the escrow peers can invoke the API *Gen_EA_KeyPair*() in the Escrow chaincode to generate the escrow agent's key pair and store the public key under the EAPK *key*.

- **Integrity and secrecy of escrowed session key under the ESS *key*.** To protect the integrity of the escrowed shared secrets, we again check the identity of the `Escrow` chaincode invoker by using the *getCreator()* API in order to make sure that only the escrow peers can invoke the `Escrow` chaincode to recover the shared secrets. On the other hand, to preserve the secrecy of the escrowed session key, we utilize on-chain private database (i.e., PDC [31]) for the ESS *keys* to prevent any unexpected peer in other organizations (i.e., data sharing organization and other key escrow organization) from reading escrowed shared secrets under the *key* that is supposed to be further used by the supervisor peer.
- **Integrity of chaincodes.** To prevent any modification to or substitution of the installed chaincodes by malicious peers, it is needed to pre-record all the version numbers of the chaincodes, and check the version numbers during the execution of PQ-KES4Chain in client codes, which developers use to invoke the chaincodes. If a mismatch of the chaincode version numbers is detected, then the chaincodes invoking and the execution of PQ-KES4Chain should be ceased.

To sum up, the execution of PQ-KES4Chain heavily relies on the characteristic of Hyperledger Fabric, which records all modification/update history of the chaincodes and on-chain data. If any peer tries to modify or counterfeit the chaincodes or related keys, then other peers can detect the modifications by checking relevant records.

4.3 Post-quantum Security Analysis

Since mainly based on smart contracts, PQ-KES4Chain only proposes a chaincode-level PQ solution of key escrow system based on Hyperledger Fabric. To protect the underlying Hyperledger Fabric from attacks from quantum computers, we suggest that developers should consider deploying PQFabric [33], which proposes a redesign of the credential-management procedures and related specifications of Hyperledger Fabric in order to incorporate hybrid digital signatures by using two signature schemes that include the quantum-safe signatures from the upcoming NIST standards.

Moreover, as one may notice, some KEM algorithms (e.g., FrodoKEM-640-AES, FrodoKEM-976-AES, SIKE-p434, SIKE-p503 and SIKE-p610) with low claimed NIST security levels from 1–3 only offer 128-bit or 192-bit shared secret that can be generated from one encapsulation-decapsulation session. Since the shared secret is supposed to be used as the AES key, one AES encryption/decryption operation with a 128-bit or 192-bit key cannot provide enough security level under the Grover attack [34] from quantum computer. This means that developers must encapsulate/decapsulate twice to generate a 256-bit shared secret and use it as one AES key to guarantee enough security level for further symmetric encryption operations.

5 Performance Evaluation

In this section, we evaluate the performance of PQ-KES4Chain, which includes the consumed execution time and the needed on-chain storage space. During the evaluation, we also highlight the performance of all the PQ KEM algorithms.

We implement PQ-KES4Chain on top of the Hyperledger Fabric version 2.2.1, the chaincodes are written in Go (1.14.13) and the client codes are written in NodeJS (12.15.0). We use AES as the symmetric algorithm and utilize the *liboqs* 0.4.0 library [35] together with its Go wrapper [36] to generate public/private key pair, encapsulate and decapsulate the shared secret for all the related PQ KEM algorithms. Since the *liboqs* library is incompatible with the native Hyperledger Fabric docker image for chaincode execution, we build a new docker image integrated with *liboqs* based on Ubuntu 18.04 and use it as the execution environment of our chaincodes. The versions of docker and docker-compose we use are 20.10.2 and 1.26.2. All the experiments are conducted on Ubuntu 18.04 VMs with 8 CPU cores of Intel Xeon E5-2680 v4 throttled to 2.4GHz and 16GB memory, and we start five VMs for the five peers (i.e., the sender, receiver, key escrow agent 1/2 and supervisor peer). For developers, all the codes, configuration files, the docker file to generate the new docker image, and related helper documents are available on [32].

5.1   Execution Time

Firstly, we test the execution time of every step in PQ-KES4Chain (except the second step because different developer may store their secret data off-chain and read the data in different ways) based on all the different PQ KEM algorithms in the NIST call. To provide enough security level, the length of pre-negotiated AES session key *SK* between sender/receiver peers is set to 256 bits, whereas the length of the secret data *M* is set to the input block length (i.e., 16 bytes) of AES algorithm.

For quick understanding, we summarize all the execution time in Fig. 2, and the number following each KEM algorithm name denotes the claimed NIST security level that the KEM algorithm can achieve. As one may notice, we exclude *Classic McEliece* algorithm from Fig. 2 because it causes much more execution time than other algorithms need and makes many data in Fig. 2 too tiny to be read. As shown in Fig. 2, the execution time of most steps in PQ-KES4Chain are within 20 or even a few milliseconds and therefore acceptable to a blockchain-based key escrow system.

We record all the execution time in Table 3, and highlight the consumed time of one key pair generation, one encapsulation operation and one decapsulation operation of each post-quantum KEM algorithm in boldface.

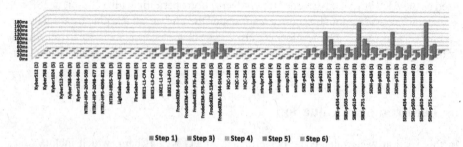

Fig. 3. The execution time of every step (except the second step) in PQ-KES4Chain based on different PQ KEM algorithms in the NIST call.

5.2   On-Chain Storage Space

We calculate the on-chain storage space needed by PQ-KES4Chain based on all the different PQ KEM algorithms. The on-chain storage space consists of three types of data tuples namely SD (Sender Data), EAPK (Escrow Agent's Public Key) and ESS (Escrowed Shared Secret) under *keys* with the same name. And one execution of PQ-KES4Chain needs one SD tuple, two EAPK tuples and two ESS tuples. As mentioned in Sect. 3.1, one SD (Sender Data) tuple contains two ciphertexts from AES encryption and two ciphertexts from PQ KEM encapsulation.

Again, for quick understanding, we summarize the on-chain storage space of PQ-KES4Chain based on different PQ KEM algorithms in Fig. 3. As one may notice in Fig. 3, we also exclude *Classic McEliece* algorithm, which causes enormous size of public key and makes the on-chain sizes based on other algorithms almost unreadable. As mentioned in Sect. 2.3, compared to the small size of shared secret, most PQ KEM algorithms have relatively big ciphertext sizes and public key sizes, which make PQ-KES4Chain storage-consuming but still acceptable to the key escrow system (Fig. 4).

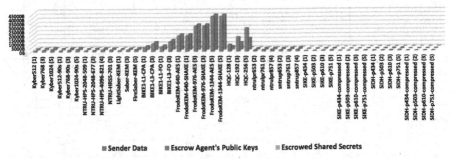

Fig. 4. The on-chain storage space of PQ-KES4Chain based on different PQ KEM algorithms (except Classic McEliece) in the NIST call.

A full list of the on-chain storage space sizes can also be found in Table 3, and we highlight the needed on-chain space of two ciphertexts, two public keys and two shared secrets of each post-quantum KEM algorithm in boldface.

5.3   Discussion About the Implications of Experiment Results

Based on the experiment results, we discuss and summarize the implications of our findings, which could be helpful for the PQ KEM application developers and the developers of PQ KEM algorithms.

- **For PQ application developers.** Intuitively speaking, the system PQ-KES4Chain based on CRYSTALS-KYBER, NTRU, SABER, NTRU Prime and HQC provide almost equally good execution time, but HQC provides 128 bytes of shared secret per KEM session, which is twice the size of shared secret most other algorithms can provide and thus could reduce the needed KEM sessions in one complex system

Table 3. The execution time (in MS) and on-chain storage (in bytes) of PQ-KES4Chain based on post-quantum KEM algorithms.

Used Algorithms (Claimed NIST security level)	Step 1) time (PQ keys gen.)	Step 3) time (PQ encap.)	Step 4) time	Step 5) time (PQ decap.)	Step 6) time	SD size (PQ cipher-texts)	EAPK sizes	ESS sizes
Classic-McEliece-348864 (1)	161 (159)	18.5 (0.3)	1.6	2.6 (0.3)	2.7	288 (256)	522240	64
Classic-McEliece-348864f (1)	133 (130.7)	16.3 (0.3)	1.4	3.1 (0.4)	2.9	288 (256)	522240	64
Classic-McEliece-460896 (3)	446 (439.5)	29.8 (0.5)	1.2	3.3 (0.6)	3	408 (376)	1048320	64
Classic-McEliece-460896f (3)	439.4 (434.2)	28.2 (0.5)	1.3	3.4 (0.6)	2.6	408 (376)	1048320	64
Classic-McEliece-6688128 (5)	679.1 (670.4)	56.9 (1)	1.4	3.4 (0.7)	2.4	512 (480)	2089984	64
Classic-McEliece-6688128f (5)	610.9 (603.5)	54.9 (1)	1.5	3.6 (0.7)	2.7	512 (480)	2089984	64
Classic-McEliece-6960119 (5)	652.8 (645.4)	49.8 (1)	1.4	3.3 (0.7)	2.7	484 (452)	2094638	64
Classic-McEliece-6960119f (5)	571.4 (562.7)	53.8 (1)	1.3	3.5 (0.7)	2.4	484 (452)	2094638	64
Classic-McEliece-8192128 (5)	690 (680.5)	64.8 (1.2)	1.6	3.6 (0.7)	2.5	512 (480)	2715648	64
Classic-McEliece-8192128f (5)	620.5 (609.5)	68.2 (1.1)	1.8	3.6 (0.7)	2.7	512 (480)	2715648	64
Kyber512 (1)	0.7 (0.1)	3.4 (0.1)	1.4	2.6 (0.1)	2.6	1568 (1536)	1600	64
Kyber768 (3)	0.7 (0.2)	3.3 (0.2)	1.7	2.4 (0.2)	2.5	2208 (2176)	2368	64
Kyber1024 (5)	0.8 (0.2)	3.7 (0.2)	1.7	2.9 (0.2)	2.7	3168 (3136)	3136	64
Kyber512-90s (1)	0.8 (0.3)	3.3 (0.2)	1.3	2.5 (0.2)	2.3	1568 (1536)	1600	64
Kyber768-90s (3)	0.7 (0.2)	3.4 (0.2)	1.7	2.5 (0.2)	2.6	2208 (2176)	2368	64
Kyber1024-90s (5)	0.9 (0.3)	3.7 (0.3)	1.8	2.8 (0.3)	2.6	3168 (3136)	3136	64
NTRU-HPS-2048-509 (1)	1.9 (1.4)	3.1 (0.1)	1.4	2.5 (0.1)	2.9	1430 (1398)	1398	64
NTRU-HPS-2048-677 (3)	2.8 (2.2)	3.5 (0.2)	1.5	2.7 (0.1)	3.1	1892 (1860)	1860	64
NTRU-HPS-4096-821 (4)	3.8 (3.1)	3.6 (0.2)	1.7	1.8 (0.2)	2.7	2492 (2460)	2460	64
NTRU-HRSS-701 (3)	2.8 (2.4)	3.2 (0.1)	1.4	2.6 (0.2)	2.8	2308 (2276)	2276	64
LightSaber-KEM (1)	0.6 (0.1)	2.9 (0.1)	1.7	2.5 (0.1)	2.5	1504 (1472)	1344	64
Saber-KEM (3)	0.8 (0.1)	3.2 (0.1)	1.8	2.8 (0.1)	2.4	2208 (2176)	1984	64
FireSaber-KEM (5)	0.7 (0.2)	3.4 (0.2)	1.4	2.5 (0.2)	2.6	2976 (2944)	2624	64
BIKE1-L1-CPA (1)	1.7 (1)	5 (1.1)	1.5	12.6 (7.9)	2.8	5116 (5084)	5084	64
BIKE1-L3-CPA (3)	4.1 (3.3)	9.9 (3.5)	1.8	34.8 (31.3)	2.8	9960 (9928)	9928	64
BIKE1-L1-FO (1)	1.8 (1.3)	5.4 (1.3)	1.4	23.3 (10.5)	2.7	5924 (5892)	5892	64
BIKE1-L3-FO (3)	5.6 (4.9)	14.1 (5.4)	2	53.5 (49.8)	2.9	12444 (12412)	12412	64
FrodoKEM-640-AES (1)	7.9 (7.2)	19.1 (7.4)	2	11.5 (7.5)	3	19472 (19440)	19232	32
FrodoKEM-640-SHAKE (1)	3.7 (3)	10 (3.3)	2.1	7.7 (3.5)	3.3	19472 (19440)	19232	32
FrodoKEM-976-AES (3)	17.2 (16.4)	37.8 (16.9)	2.3	21.7 (16.9)	3.5	31520 (31488)	31264	48
FrodoKEM-976-SHAKE (3)	7.4 (6.5)	18.5 (7.2)	2.5	11.6 (7.2)	3.6	31520 (31488)	31264	48
FrodoKEM-1344-AES (5)	31.6 (30.6)	68.2 (31.7)	2.9	37.2 (31.7)	4	43296 (43264)	43040	64
FrodoKEM-1344-SHAKE (5)	12.5 (11.6)	30.3 (12.8)	2.7	18.6 (12.9)	4	43296 (43264)	43040	64
HQC-128 (1)	0.9 (0.3)	3.7 (0.4)	1.7	3.8 (0.7)	3	8994 (8962)	4498	128
HQC-192 (3)	1.6 (0.7)	5.3 (1.1)	2.1	4.9 (1.5)	3.1	18084 (18052)	9044	128
HQC-256 (5)	1.7 (1.1)	7.7 (2.2)	2.4	7 (3)	3.4	28970 (28938)	14490	128
ntrulpr653 (2)	1 (0.5)	4.7 (0.8)	2	3.6 (1)	2.6	2082 (2050)	1794	64
ntrulpr761 (3)	1.2 (0.6)	4.8 (1)	1.7	3.7 (1.5)	2.5	2366 (2334)	2078	64
ntrulpr857 (4)	1.3 (0.8)	5.4 (1.2)	1.8	4.2 (1.6)	2.4	2656 (2624)	2368	64
sntrup653 (2)	6.6 (6)	3.6 (0.4)	1.9	3.3 (1)	2.4	1826 (1794)	1988	64
sntrup761 (3)	8.2 (7.7)	3.9 (0.5)	1.8	3.7 (1.3)	2.6	2110 (2078)	2316	64
sntrup857 (4)	11.1 (10.6)	4.9 (0.7)	1.9	3.9 (1.5)	2.5	2400 (2368)	2644	64
SIKE-p434 (1)	11.6 (11.1)	38.7 (17.9)	1.7	21.6 (19.2)	2.9	724 (692)	660	32
SIKE-p503 (2)	4.5 (3.9)	15.7 (6.3)	1.5	9.2 (6.7)	2.7	836 (804)	756	48
SIKE-p610 (3)	32.9 (32.4)	122.3 (59.6)	1.8	62.5 (60)	2.6	1004 (972)	924	48
SIKE-p751 (5)	12.5 (12.1)	41 (19)	1.4	23.1 (20.8)	2.6	1224 (1192)	1128	64
SIKE-p434-compressed (1)	20.8 (20.4)	59.1 (28.1)	1.4	23.2 (20.7)	2.3	504 (472)	394	32
SIKE-p503-compressed (2)	7.5 (6.9)	22.9 (10.1)	1.6	9.7 (7.4)	2.3	592 (560)	450	48
SIKE-p610-compressed (3)	55.6 (54.9)	164.9 (81.1)	1.4	65.5 (63.4)	2.4	704 (672)	548	48
SIKE-p751-compressed (5)	20.6 (20.1)	65.2 (30.9)	1.5	24.5 (22.2)	2.2	852 (820)	670	64
SIDH-p434 (1)	10.6 (10.1)	44.2 (20.3)	1.6	10.5 (8.1)	2.9	692 (660)	660	220
SIDH-p503 (2)	4 (3.6)	17.1 (7.2)	1.3	5.2 (2.8)	2.4	788 (756)	756	252
SIDH-p610 (3)	33 (32.5)	121.7 (59.4)	1.5	29.5 (27.3)	2.4	956 (924)	924	308
SIDH-p751 (5)	11 (10.6)	46.9 (21.9)	1.4	11.1 (8.8)	2.3	1160 (1128)	1128	376
SIDH-p434-compressed (1)	20.4 (19.9)	59.1 (28.1)	1.3	11.7 (9.4)	2.3	426 (394)	394	220
SIDH-p503-compressed (2)	7.2 (6.8)	23 (10.1)	1.2	5.7 (3.5)	2.5	482 (450)	450	252
SIDH-p610-compressed (3)	55.3 (54.8)	166.5 (81.8)	1.3	32.5 (30.1)	2.4	580 (548)	548	308
SIDH-p751-compressed (5)	19.9 (19.4)	63.9 (30.5)	1.4	11.9 (9.8)	2.7	702 (670)	670	376

and further save the total execution time of the complex system. On the other hand, the on-chain storage space based on SIKE is the smallest, followed by CRYSTALS-KYBER, NTRU, SABER and NTRU Prime. To sum up, the PQ KEM application developers should choose HQC (resp. SIKE) if the application is only time-sensitive (resp. storage-sensitive), while CRYSTALS-KYBER, NTRU, SABER and NTRU Prime provide equally good overall performance on both execution time and on-chain storage.

- **For PQ KEM algorithm developers.** Based on the characteristics of blockchain-based applications, we make some suggestions for the PQ KEM algorithm developers to help them improve their algorithms to meet the requirements of the blockchain-based systems and be more competitive in the NIST call. (1) *Longer shared secret length is more favorable.* Since the shared secret (normally 128–256 bytes) is further used as the symmetric key (e.g., AES key) to encrypt/decrypt on-chain data, to provide enough post-quantum security level, the length of the symmetric key (i.e., the shared secret) must meet corresponding requirements (e.g., 256-bit AES key). Therefore, to achieve the same post-quantum security level, a longer shared secret length can efficiently reduce the number of KEM sessions and related execution time/on-chain storage space in the system. (2) *Private key size could be designed longer for better performance.* Normally, private key is not stored on the public ledger, so it has no impact on the on-chain storage. The PQ algorithm developers could upgrade their algorithms by increasing the private key size in return for shorter public key size, shorter encapsulated shared secret size, longer shared secret length or better execution time.

6 Conclusion

In this paper, we proposed PQ-KES4Chain, which is the first post-quantum key escrow system for consortium blockchain guaranteeing both data confidentiality and supervised data disclosure under the threat from quantum computers. In PQ-KES4Chain, we integrated the system with all the PQ public-key encryption/KEM algorithms in the current round of NIST call for national standard. We implemented the PQ-KES4Chain system on top of Hyperledger Fabric, and provided chaincodes, related APIs together with client codes for further development. Finally, we conducted security analysis and performance evaluation on PQ-KES4Chain. Based on the experiment results, we discussed and summarized the implications of our findings, which could be helpful for the developers of the PQ KEM algorithms and applications.

Acknowledgment. This research is supported by the National Key R&D Program of China (Program No. 2018YFC604005), the National Natural Science Foundation of China (Grant No. 61902285, 62004077), and the Fundamental Research Funds for the Central Universities (Program No. 2662017QD041, 2662018QD043). We also thank Pawel Szalachowski for the valuable discussions and feedback.

References

1. Hyperledger. https://www.hyperledger.org/use/fabric. Accessed 01 Sept 2021
2. Quorum. https://www.goquorum.com. Accessed 01 Sept 2021
3. Dorothy, E.D., Dennis, K.B.: A taxonomy for key escrow encryption systems. Commun. ACM **39**(3), 34–40 (1996)
4. Peter, W.S.: Polynomial-time algorithms for prime factorization and discrete logarithms on a quantum computer. SIAM J. Comput. **26**(5), 1494–1509 (1997)
5. Post-Quantum Cryptography: Round 3 Submissions. https://csrc.nist.gov/Projects/post-quantum-cryptography/round-3-submissions. Accessed 01 Sept 2021

6. Matt, B.: Protocol failure in the escrowed encryption standard. In: Proceedings of ACM CCS, Fairfax, Virginia, USA, pp. 59–67 (2010)
7. David, M.B., Carl, M.E., Steven, B.L., Stephen, T.W.: A new approach to software key escrow encryption (1994)
8. Dawson, E., He, J.: Another approach to software key escrow encryption. In: Pieprzyk, J., Seberry, J. (eds.) ACISP 1996. LNCS, vol. 1172, pp. 87–95. Springer, Heidelberg (1996). https://doi.org/10.1007/BFb0023290
9. Stephen, T.W., Steven, B.L., Carl, M.E., David, M.B.: Commercial key recovery. Commun. ACM **39**(3), 41–47 (1996)
10. Adi, S.: Partial key escrow: a new approach to software key escrow. Presented at Key Escrow Conference, Washington, D.C., 15 September 1995
11. Mihir, B., Shafi, G.: Verifiable partial key escrow. In: Proceedings of ACM CCS, Zurich, Switzerland, pp. 78–91 (1997)
12. Mihir, B., Shafi, G.: Encapsulated key escrow. Computer science technical report 688. MIT Laboratory (1996)
13. Goldfeder, S., Bonneau, J., Gennaro, R., Narayanan, A.: Escrow protocols for cryptocurrencies: how to buy physical goods using bitcoin. In: Kiayias, A. (ed.) FC 2017. LNCS, vol. 10322, pp. 321–339. Springer, Cham (2017). https://doi.org/10.1007/978-3-319-70972-7_18
14. Sungyong, C., Seungsoo, B., Seungjoo, K.: Blockchain based sensitive data management by using key escrow encryption system from the perspective of supply chain. IEEE Access **8**, 154269–154280 (2020)
15. Wenzhe, L., Sheng, W., Chunxiao, J., et al.: Towards large-scale and privacy-preserving contact tracing in COVID-19 pandemic: a blockchain perspective. IEEE Trans. Netw. Sci. Eng. (2020). Early Access
16. Robert, C.: Transitioning to a hyperledger fabric quantum-resistant classical hybrid public key infrastructure. J. Br. Blockchain Assoc. **2**(2), 1–11 (2019)
17. Robert, C.: Evaluation of post-quantum distributed ledger cryptography. J. Br. Blockchain Assoc. **2**(1), 1–8 (2019)
18. Shen, R., Xiang, H., Zhang, X., Cai, B., Xiang, T.: Application and implementation of multivariate public key cryptosystem in blockchain (short paper). In: Wang, X., Gao, H., Iqbal, M., Min, G. (eds.) CollaborateCom 2019. LNICSSITE, vol. 292, pp. 419–428. Springer, Cham (2019). https://doi.org/10.1007/978-3-030-30146-0_29
19. Semmouni, M.C., Nitaj, A., Belkasmi, M.: Bitcoin security with post quantum cryptography. In: Atig, M.F., Schwarzmann, A.A. (eds.) NETYS 2019. LNCS, vol. 11704, pp. 281–288. Springer, Cham (2019). https://doi.org/10.1007/978-3-030-31277-0_19
20. Tiago, M.F.-C., Paula, F.-L.: Towards post-quantum blockchain: a review on blockchain cryptography resistant to quantum computing attacks. IEEE Access. **8**, 21091–21116 (2020)
21. Micali, S.: Fair public-key cryptosystems. In: Brickell, E.F. (ed.) CRYPTO 1992. LNCS, vol. 740, pp. 113–138. Springer, Heidelberg (1993). https://doi.org/10.1007/3-540-48071-4_9
22. Classical McEliece. https://classic.mceliece.org. Accessed 01 Sept 2021
23. Kyber. https://pq-crystals.org/kyber/index.shtml. Accessed 01 Sept 2021
24. NTRU. https://ntru.org/. Accessed 01 Sept 2021
25. SABER. https://www.esat.kuleuven.be/cosic/pqcrypto/saber/. Accessed 01 Sept 2021
26. BIKE. https://bikesuite.org. Accessed 01 Sept 2021
27. NTRU Prime. https://ntruprime.cr.yp.to. Accessed 01 Sept 2021
28. FrodoKEM. https://frodokem.org. Accessed 01 Sept 2021
29. SIKE. https://sike.org. accessed 2021/09/01
30. HQC, https://pqc-hqc.org. Accessed 01 Sept 2021
31. Hyperledger Fabric private data. https://hyperledger-fabric.readthedocs.io/en/release-2.2/private-data/private-data.html. Accessed 01 Sept 2021

32. PQ-KES4Chain. https://github.com/conf-auth/PQ-KES4Chain. Accessed 01 Sept 2021
33. Holcomb, A., Pereira, G., Das, B., Mosca, M.: PQFabric: a permissioned blockchain secure from both classical and quantum attacks. In: Proceedings of IEEE International Conference on Blockchain and Cryptocurrency, Virtual Conference, pp. 1–9 (2021)
34. Lov, K.G.: A fast quantum mechanical algorithm for database search. In: Proceedings of STOC, New York, NY, USA, pp. 212–219 (1996)
35. Liboqs. https://github.com/open-quantum-safe/liboqs. Accessed 01 Sept 2021
36. Liboqs-go. https://github.com/open-quantum-safe/liboqs-go. Accessed 01 Sept 2021

Flexible and Survivable Single Sign-On

Federico Magnanini[1]([✉]), Luca Ferretti[1], and Michele Colajanni[2]

[1] University of Modena and Reggio Emilia, Modena, Italy
{federico.magnanini,luca.ferretti}@unimore.it
[2] University of Bologna, Bologna, Italy
michele.colajanni@unibo.it

Abstract. Single sign-on (SSO) is a popular authentication method that is vulnerable to attacks exploiting the single points of failure of its centralized design. This problem is addressed by survivable SSO protocols relying on distributed architectures that enable a set of servers to collectively authenticate a user. However, existing survivable SSO protocols have limitations because they do not allow service providers to modify security parameters after protocol setup. This paper introduces the first survivable SSO protocol that guarantees flexibility. This property is of utmost importance for SSO because it allows service providers to tailor the trade-off between performance overhead and security requirements of multiple services and even to preserve compatibility with non-survivable SSO.

Keywords: Survivability · Single sign-on · Distributed systems · Cloud · Flexibility

1 Introduction

Single sign-on (SSO) is a popular protocol to authenticate users requiring access to multiple Web-based services. Typically, an identity provider manages one logical identity server that issues authentication tokens proving user identity to service providers. The centralized design of SSO protocols is vulnerable to authentication tokens forgery as demonstrated by recent incidents [2,10] where cyber attackers compromised the identity server and forged tokens that falsely impersonated users towards service providers. This issue can be addressed by so-called *survivable SSO protocols* that can prevent user impersonation even in presence of attacks (e.g., [3,6,26]). In survivable SSO, the identity provider manages multiple identity servers. A user authenticates himself to a subset of identity servers that collectively sign an authentication token and demonstrate the user identity to service providers. The amount of signing identity servers must be greater than a security threshold which defines the maximum number of identity servers that the adversary can violate.

Existing proposals achieve survivable token release by signing tokens through threshold signatures, and survivable user authentication through distributed password-based authentication protocols [3,6,26]. The problem is that threshold signatures tend to be unrealistic in practice because they do not guarantee

© Springer Nature Switzerland AG 2022
W. Meng and M. Conti (Eds.): CSS 2021, LNCS 13172, pp. 182–197, 2022.
https://doi.org/10.1007/978-3-030-94029-4_13

flexibility. They prevent service providers from dynamically adjusting the value of the security threshold during protocol execution and they are not backwards-compatible with existing SSO systems. The lack of flexibility and of backwards compatibility prevent service providers from offering services with different identity assurance levels [1,16], and from the possibility of dynamically adjusting the threshold based on user contextual information as suggested by the recent zero trust paradigm [25].

We propose an original survivable SSO protocol where survivable token release is achieved by signing authentication tokens through conventional digital signatures instead of threshold signatures as in literature. This approach enables the design of a survivable token release scheme that guarantees flexibility and preserves backwards compatibility with non-survivable SSO solutions. Moreover, we show that it is possible to guarantee survivable user authentication through password-based protocols even if they are not designed for distributed architectures. We evaluate the security of the proposed token release scheme and show the security of the overall SSO protocol by considering existing password-based authentication methods.

The practical relevance of the proposal is twofold. First, flexible and survivable SSO can be leveraged to mitigate emerging attacks to access confidential cloud resources through rogue authentication tokens [10]. Moreover, the high security level that is guaranteed by flexible and survivable SSO protocols is a perfect combination for mission critical systems requiring robust and reliable authentication mechanisms even under attack [13].

The paper is organized as follows. Section 2 discusses related work. Section 3 describes the system model and the single sign-on framework. Section 4 describes the threat model. Section 5 discusses the details and guarantees of the proposed novel survivable token release scheme. Section 6 shows the security level of the proposal. Section 7 evaluates the security and flexibility of the proposal and of related SSO protocols when integrated with different credential management systems. Section 8 highlights final remarks.

2   Related Work

This paper is related to recent results investigating SSO protocols with different trade-offs between survivable security guarantees and flexible configuration [3,6,26]. For example, the authors in [3] propose a SSO protocol that tolerates the violation of up to a threshold of identity servers. The value of the threshold can be configured at setup time to tolerate from one compromised identity server up to a dishonest majority including a single honest identity server. The problem is that this protocol does not guarantee recoverability because it does not include the due procedures to recover compromised identity servers to a safe state. As a result, the protocol cannot be considered survivable because recoverability is a mandatory security guarantee in these systems [5]. We propose a protocol that guarantees survivability by defining specific procedures to recover compromised identity servers.

The guarantee of survivability is also analyzed by the proposal of a password-based survivable SSO protocol [6]. The authors consider a strong adversarial model that requires to trade token unforgeability for availability, as the protocol does not terminate if a single identity server is unavailable. We assume a different model called *mobile adversarial model* [17]. Although it is weaker than that considered in [6], the mobile adversary is a realistic model for SSO scenarios and allows the proposed protocol to guarantee termination even in presence of a fully malicious minority of identity servers. The authors of [6] achieve survivable token release by signing authentication tokens through an original RSA-based threshold signature scheme. However, the lack of flexibility of threshold signatures represents a major problem as they force the identity provider to set the value of the security threshold at setup time. Moreover, threshold signatures cause management issues as they cannot guarantee backwards compatibility with non-survivable SSO protocols.

The proposed protocol guarantees flexible SSO and offers the possibility of choosing the value of the security threshold at verification time. This allows a service provider to offer multiple services with different identity assurance levels (e.g., [1,16]) and to choose the most suitable threshold for each of them. Moreover, it is possible to dynamically adjust the security threshold depending on user contextual information, as suggested by the zero trust paradigm [25], and to tailor the best trade-off between performance and security for each service. Finally, the proposed flexible and survivable SSO can preserve compatibility with non-survivable SSO. This would enable a gradual transition of existing service providers towards survivable SSO. A service provider can enable backwards compatible support to survivable SSO by updating the authentication token verification algorithm.

The password-based survivable SSO proposed in [26] obtains a better trade-off in terms of threshold configuration and survivability with respect to [3] and [6]. It allows the configuration of the security threshold at setup time and guarantees survivability. Although the authors do not explicitly specify an adversarial model, their proposal seems to consider a mobile adversary similar to that proposed in this paper. While their protocol obtains good trade-offs, it lacks flexibility due to the adoption of threshold signatures during the token authentication phase.

3   System Model

We describe the survivable SSO protocol by referring to Fig. 1 showing the main entities, data and operation flows. The proposed protocol involves four entities: *user*, *service provider*, *identity provider*, and a set of *identity servers*. The user denotes a person who wants to access services and resources maintained by the service provider. The identity provider denotes an authority that defines and operates a set of identity servers to offer the survivable single sign-on protocol. The protocol involves the following types of data:

- *user credentials*: unique information held by each user presented to identity servers for authentication;
- *credentials databases*: data structures independently maintained by identity servers to verify users credentials;
- *partial tokens*: assertions about users identities authenticated by a single identity server;
- *authentication tokens*: assertions about users identities whose authenticity is guaranteed by a subset of identity servers;
- *token signing keys*: secret cryptographic material held by each identity server to authenticate authentication tokens;
- *identity provider certificate*: public cryptographic material used by the service provider to verify authentication tokens;
- *identity provider signing key*: secret cryptographic material used by the identity provider to authenticate the identity provider certificate.

The proposed protocol consists of five operations.

- *Setup*: the identity provider defines the initial set of identity servers, and releases a certificate that authenticates identity servers public keys. We assume that the certificate is distributed to all actors by using orthogonal public key distribution protocols [8];
- *Register*: the user registers his credentials to all identity servers credentials databases;
- *Sign-on*: the user requests an authentication token to identity servers. This operation is composed by the following steps:
  • *Verify credential*: an identity server verifies user credentials against his credential database;

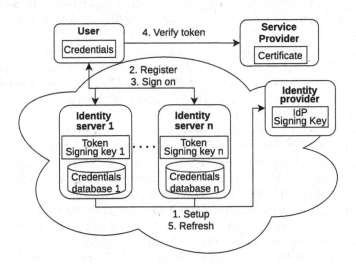

Fig. 1. Architecture and high level protocol flow

- *Release partial*: an identity server releases a partial token to an authenticated user;
- *Combine*: the user combines the partial tokens collected by a threshold of identity servers in an authentication token;
- *Verify token*: the service provider uses the identity provider certificate to verify the authenticity and validity of the authentication token presented by the user;
- *Refresh*: the identity provider updates identity servers secret cryptographic material.

The protocol requires that users establish secure (confidential and authenticated) bidirectional communication channels with legitimate identity servers by using public keys of identity servers distributed during the *Setup* phase. Communication channels allow users to authenticate identity servers, and not vice versa, as it is common for standard HTTPS communications.

The proposed framework represents a novel contribution to model the operations of survivable single sign-on protocols. It captures common operations shared by related proposals which were not highlighted by previous literature, and allows us to compare the proposed protocol with related works [3, 6, 26]. The framework extends existing non-survivable single sign-on frameworks by introducing the Release partial and Combine procedures. Existing related survivable SSO protocols adopt a similar system model where identity servers are coordinated through decentralized protocols which do not require an identity provider [3, 6, 26]. However, they do not guarantee flexibility (see Sect. 6). The proposed architecture introduces the additional role of the identity provider because the considered scenarios, such as cloud-based SSO, are characterized by centralized governance where the identity provider acts as an authority that operates identity servers during Setup and Refresh operations. At Setup time, the identity provider defines the infrastructure while at Refresh time it proactively secures it.

The proposal does not limit the identity provider from being a decentralized entity because it can be extended to support decentralized execution of the Setup and Refresh operations by leveraging ideas from [21]. For ease of presentation and without loss of generality, in the remainder of this paper we consider the identity provider as one entity.

We enable identity providers to offer flexible and survivable SSO as a service. An identity provider can execute the Setup operation by defining the maximum security threshold k_{max} and by provisioning an infrastructure of $2k_{max}+1$ failure-independent identity servers. Service providers can choose the most appropriate security threshold value between zero and k_{max} to enforce survivability on their services. A security threshold equal to zero maintains compatibility with existing non-survivable SSO. A security threshold equal to k_{max} allows service providers to enforce the highest level of identity assurance even in presence of k_{max} compromised identity servers. Guaranteeing a practical failure independence of all servers against benign and malicious faults tends to become quickly an intractable challenge as demonstrated in [15, 19]. Hence, we can assume that in practice the value of k_{max} is at most of few units. If we accept stronger security assumptions on identity servers, then the value of k_{max} can be increased above the few units. For example, some identity servers may share the same operating system. This security trade-off simplifies the technological challenge of provisioning and maintaining several operating systems to enable the deployment of a larger yet less diverse set of identity servers.

4   Threat Model

We discuss possible attacks from a twofold perspective: we discuss violation and recovery patterns throughout the protocol lifetime; we analyze the multiple classes of attacks that an adversary can adopt to subvert the protocol security. We use these analyses to assess the security guarantees of the proposal in Sects. 6 and 7.

For the threat analysis, we consider the popular *mobile adversary model* that aims at subverting the survivable single sign-on protocol. This model was proposed in the context of distributed function evaluation [23] and applied to secret sharing [17], multiparty computation [12], intrusion tolerant certification authorities [27], key management systems [11], cloud-based secure logging [24] and secure software update systems [20]. It assumes that identity servers are failure-independent and that at any instant an identity server is either *honest* or *compromised*. A compromised identity server can be recovered and become honest after that its hardware and software have been reset to a known clean state and its secret cryptographic material has been obsoleted. A recovery of all identity servers is periodically executed by the identity provider during the Refresh operation. The proposed protocol tolerates that a minority of identity servers is compromised simultaneously, that is, it guarantees security against adversaries that violate up to $k < n/2$ identity servers, where n is the total number of identity servers. This is the typical security level guaranteed by related works using the mobile adversary model (e.g., [11,17]).

We discuss violation and recovery patterns between periodic Refresh operations by referring to Fig. 2. The time horizon is divided in *time periods*, where each begins with the execution of the Refresh operation. We denote the remaining part of the time period after the completion of the Refresh as *operation period*. During the operation period, honest identity servers operate the single sign-on protocol according to initial specification. Each highlighted area represents a time interval during which the adversary has compromised up to k identity servers.

The mobile adversary model considers four attack patterns. The first pattern captures an adversary that has corrupted up to k identity servers during the operation period and that is removed by the identity provider during the next Refresh operation. The second pattern considers an adversary that has corrupted up to k identity servers during the Refresh operation. The adversary has the additional power to interfere with the identity provider during Refresh. The third and fourth patterns capture a powerful and elusive adversary that is able

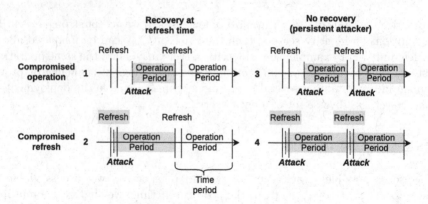

Fig. 2. Violation patterns

to move laterally among identity servers and ensures a persistent presence at the identity provider infrastructure even after Refresh operations. An attacker cannot control more than k identity servers during each period.

An adversary can perform different types of attacks to subvert the survivable single sign-on protocol. We label attacks to reference each of them when analyzing security guarantees in Sects. 6 and 7.

- **A.** Violation of the token release system within identity servers:
  - **A1:** the adversary violates the confidentiality of the token signing keys within identity servers to forge authentication tokens;
  - **A2:** the adversary violates the integrity of the token release system by returning bogus partial tokens to users;
  - **A3:** the adversary violates the availability of the token release system by deleting token signing keys or by not returning partial tokens to users.
- **B.** Violation of the identity verification protocols within identity servers:
  - **B1:** the adversary compromises the integrity of the credentials database of an identity server to forcefully set known credentials to users accounts;
  - **B2:** the adversary violates the availability of the identity verification protocol by deleting the credential database or by not completing credential verification;
  - **B3:** the adversary violates the confidentiality of the credentials database of an identity server to recover users credentials;
  - **B4:** the adversary reads an identity server internal state during credential verification to recover users credentials;
  - **B5:** the adversary executes Man-in-The-Middle attacks from compromised identity servers to impersonate the user at honest identity servers during the Sign-on operation.
- **C.** Violation of the identity provider:
  - **C1:** the adversary violates the confidentiality of the identity provider signing key.

5 Survivable Token Release

A comprehensive analysis of the proposed SSO protocol is in Sect. 7. Here, we focus on the token release scheme. The key insight of the proposed token release scheme is twofold. First, signing authentication tokens through conventional digital signatures allows to achieve flexibility. Second, proactively rotating token signing keys allows to guarantee security against lateral movement of the mobile adversary while the identity provider allows to efficiently authenticate rotated public keys and modified system parameters. We consider a black box digital signature scheme defined by the following operations framework:

- $\langle sk, pk \rangle \leftarrow$ KeyGen(): generate secret key sk and public key pk;
- $\sigma \leftarrow$ Sign(sk, m): compute signature σ on message m with secret key sk;
- $0 \vee 1 \leftarrow$ Verify(pk, m, σ): if signature σ authenticates message m under public key pk output 1, 0 otherwise.

We assume that the identity provider has established the security level of the adopted cryptographic schemes and that the resulting public parameters are known to all actors. For ease of notation, we omit public parameters from the scheme operations. The proposed token release scheme implements the following operations.

$crt \leftarrow$ Setup($sk_{IdP}, k_{max}, \{\langle pk_1, \pi_1 \rangle \dots, \langle pk_n, \pi_n \rangle\}$). The identity provider, given his secret key sk_{IdP} and security threshold k_{max} verifies the identity servers certificate signing requests $\{\pi_1, \dots, \pi_n\}$ and authenticates the corresponding public keys $\{pk_1, \dots, pk_n\}$ by computing the identity provider certificate crt. To this aim it executes the following steps:

- the service provider defines the value k_{max} and a set of $n = 2k_{max} + 1$ identity servers;
- each identity server executes the KeyGen() algorithm to compute the signing key pair $\langle sk_i, pk_i \rangle$;
- each identity server i then computes a certificate signing request π_i for pk_i, and sends π_i to the identity provider over a secure channel. Certificate signing requests can be computed with well-established standard algorithms [22].
- the identity provider collects the public keys $PK = \{pk_1, \dots, pk_n\}$ and verifies each certificate signing request in $\{\pi_i\}$;
- if all received certificate signing requests $\{\pi_i\}$ are valid, the identity provider authenticates the corresponding set of public keys and the value k_{max} by signing the pair $\langle PK, k_{max} \rangle$ with his private key sk_{IdP}, producing $crt = \langle PK, k_{max}, \sigma_{crt} \rangle$;
- the identity provider publicly distributes crt.

The authentication methods used in procedures Register and VerifyCredential are orthogonal to the proposed token release scheme in terms of functionality, hence we can omit details. A comprehensive discussion about integration with different user authentication methods is in Sect. 7.

$\sigma_i \leftarrow$ ReleasePartial(sk_i, id): identity server i with secret key sk_i, given identity id produces partial token σ_i. The ReleasePartial procedure assumes that identity id has been already verified during VerifyCredential. The identity server signs id with his secret key sk_i by executing algorithm $\mathsf{Sign}(sk_i, id)$ of the signature scheme. The identity server then sends the resulting signature σ_i to the user over a secure channel.

$\langle id, \{\sigma_i\}, L\rangle \leftarrow \mathsf{Combine}(id, \{\sigma_i\}_{i\in L})$: the user verifies partial tokens $\{\sigma_i\}_{i\in L}$ collected from the set of identity servers L, and outputs the authentication token $\langle id, \{\sigma_i\}, L\rangle$. To this aim he executes the following steps:

- when the user has collected $k+1$ partial tokens, as required by the service provider, the user determines the set of servers L that produced the collected partial tokens;
- the user then verifies that each of the collected partial tokens is authentic, by executing the $\mathsf{Verify}(pk_i, id, \sigma_i)$ for each $i \in L$;
- if all partial tokens are authentic, the user outputs the authentication token $\langle id, \{\sigma_i\}, L\rangle$.

$0 \vee 1 \leftarrow \mathsf{VerifyToken}(\langle id, \{\sigma_i\}, L\rangle, crt, k)$: the service provider verifies whether all signatures in $\{\sigma_i\}$, produced by the set of identity servers L, authenticate id by using crt. It outputs 0 if any σ_i does not authenticate id or if $|L| < k+1$, 1 otherwise. To this aim it executes the following steps:

- the service provider verifies the authenticity of the identity provider certificate $crt = \langle PK, k_{max}, \sigma_{crt}\rangle$ by verifying signature σ_{crt}. The result of this operation can be cached as long as the set PK is not refreshed;
- it verifies that $|L| \geq k+1$;
- it verifies all signatures in $\{\sigma_i\}$ by executing $\mathsf{Verify}(pk_i, id, \sigma_i)$;
- if any of the previous checks fails it outputs 0, otherwise it outputs 1.

$crt' \leftarrow \mathsf{Refresh}(sk_{IdP}, crt, PK_a, PK_r)$: the identity provider sets the new identity provider certificate crt', given the current identity provider certificate crt, the new set of public keys PK_a and the set of revoked public keys PK_r. To this aim he executes the following steps:

- the identity provider recovers all compromised identity servers and establishes the set of additional identity servers, if any.
- each additional or recovered identity server in the new set computes his signing key pair $\langle sk_i, pk_i\rangle$ by executing the $\mathsf{KeyGen}()$ algorithm, produces the corresponding certificate signing request π_i and sends it to the identity provider.
- the identity provider collects the set of new public keys PK_a, verifies the corresponding certificate signing requests $\{\pi_i\}_{i\in|PK_a|}$, and defines the set PK_r of public keys to revoke. PK_r includes the old public keys of recovered servers and any disposed server that is excluded from the protocol.
- the identity provider sets the new set of public keys $PK' = PK \cup PK_a \backslash PK_r$ such that $|PK'| = |PK|$, and authenticates the pair $\langle PK', k_{max}\rangle$ by signing it with his private key, producing $crt' = \langle PK', k_{max}, \sigma_{crt'}\rangle$ which is made publicly available.

6   Security Guarantees

In this section we discuss the security of the token release scheme against attack classes A and C that we presented in Sect. 4. We devote Sect. 7 to consider class B attacks and evaluate the security of the overall SSO protocol when the token release scheme is integrated with different credential management systems. Here, we show how to mitigate class C attacks that violate the confidentiality of the identity provider signing key. Moreover, we show that the proposed token release scheme is secure against class A attacks that aim to violate the token release system within identity servers. Furthermore, we show that security against class A attacks holds during all violation patterns (Fig. 2) that is, violation during: one operation period (1), one Refresh execution (2), consecutive operation periods (3) and consecutive Refresh executions (4).

A1. The attacker can violate the confidentiality of the token signing key of up to k identity servers through any of the violation patterns. The attacker can force a compromised identity server to execute the ReleasePartial operation on identities of the attacker's choice. We note that the attacker can obtain the same level of violation even if it only controls the key without knowing its value (e.g. the key is protected by an HSM). Attack A1 with violation pattern 1 is ineffective because an attacker that controls no more than k identity servers is not able to forge a valid authentication token. Attack A1 is ineffective even with the violation pattern 2. Here, the identity provider verifies that the certificate signing request submitted by identity servers is legitimate before certifying the new public keys. Moreover, an adversary that has access to the new secret keys produced during Refresh execution is unable to obtain a valid authentication token since he does not control enough $(k + 1)$ identity servers. Finally, attack A1 is also ineffective with violation patterns 3 and 4. Given that identity servers are recovered at the beginning of the Refresh procedure, the adversary can never control more than k identity servers within any time period. Therefore, the adversary does not control enough $(k + 1)$ identity servers to forge authentication tokens. Moreover, the public keys of the recovered identity servers are revoked during Refresh. We note that the attacker may try to exploit race conditions during revocation of public keys (e.g., CRL propagation delays) to force the service provider to verify authentication tokens with revoked public keys. To prevent this type of attacks we rely on the identity provider as an online certification authority. In this way the service provider always validates authentication tokens with a fresh copy of the new set of public keys PK'. As a result, the service provider can easily detect invalid authentication tokens authenticated by revoked public keys.

A2. Attack A2 is ineffective in any violation pattern because the proposed token release scheme allows the user to detect bogus partial tokens, discard them and repeat the *Sign-on* operation with another honest identity server; its existence is guaranteed by to the presence of an honest majority.

A3. Attack A3 is ineffective in any violation pattern because the Refresh operation ensures that at any moment there is an available honest majority of $k + 1$ identity servers that is able to respond to users.

C1. The attacker may try to violate the identity provider signing key. However, we note the identity provider secret key sk_{IdP} can be kept offline as it must be used only during Setup and Refresh operations which occur on a much broader frequency than operations involving the identity servers signing keys. As a result, the signing key of identity provider can be protected to ensure it less vulnerable than the signing keys of identity servers. We remind that the survivability of the proposal could be extended to the identity provider by instantiating it as a collective entity as already discussed in Sect. 3.

The proposed protocol guarantees also the *accountability* security property that is, it allows an identity provider to attribute protocol deviations to specific compromised identity servers. This property is important because it allows an identity provider to prioritize recovery operations when servers are compromised, and can be complementary to already deployed approaches for monitoring system infrastructure [4]. All cryptographic material issued by identity servers is accountable. For example, each partial token σ_i produced during the ReleasePartial procedure is accountable because it can be verified through the public key of identity server i. Moreover, the authentication token $\langle id, \{\sigma_i\}, L \rangle$ computed during the Combine procedure is accountable because the set of signers L explicitly indicates the identity servers that contributed to signatures $\{\sigma_i\}$. It is important to note that related works [3,6,26] relying on threshold signatures are not completely accountable. Partial tokens computed by identity servers during the *Release partial* procedure may be accountable depending on the adopted scheme, whereas authentication tokens are not accountable. Other papers (e.g., [3,26]) adopt the threshold signature scheme of Boldyreva [9] which allows accountability of partial tokens because identity servers publish a commitment of their secret shares after the *Setup* and *Refresh* procedures. The work in [6] proposes an original RSA-based threshold signature scheme which blinds partial tokens thus preventing partial token accountability. The authentication token computed during the *Combine* procedure of [3,6,26] is not accountable because a key property of threshold signatures is that they do not reveal the identity of individual signers but only the cardinality of the set of signers [9].

The proposed original token release scheme guarantees several flexibility benefits: adjustable security threshold, adjustable performance overhead and compatibility with non-survivable SSO, that we discuss below. The proposed token release scheme guarantees an *adjustable security threshold* because the service provider can decide the value of k during the VerifyToken operation. This allows the choice of the best trade-off between security and performance for the service he offers, provided that $0 \leq k \leq k_{max} = \lfloor \frac{n}{2} \rfloor$. The constraint $k_{max} = \lfloor \frac{n}{2} \rfloor$ guarantees availability in case of k_{max} disconnected identity servers.

The scheme also guarantees an *adjustable performance overhead* because it allows the service provider to adjust the value of k to the best trade-off between performance and security. Higher values of k imply higher security in terms of token unforgeability and availability, as the adversary is required to violate $k+1$ identity servers to issue rogue authentication tokens or impede their release. However, these higher k values imply higher overheads because a user executes the sign-on procedure with $k + 1$ servers. Previous proposals [3, 6, 26], which are based on threshold signatures, do not achieve a comparable flexibility because the value k is not decided by the service provider, but by the identity provider during the *Setup* operation. This value cannot be changed afterwards.

As a final attribute, the proposed token release scheme preserves *compatibility with non-survivable SSO*. If a service provider sets $k = 0$ during the VerifyToken operation, its users execute the SSO procedure with one identity server as in traditional non-survivable SSO.

7   Integration with Credentials Verification and Storage Protocols

We consider different credentials storage and verification protocols to evaluate their impact on the security of the SSO protocol. To this aim, we consider the attack category B described in Sect. 4, which involve credentials verification and storage protocols as follows:

- B1: integrity violation of credentials database;
- B2: unavailable identification protocol;
- B3: credentials database leak;
- B4: internal state leak;
- B5: Man-In-The-Middle (MITM) attacks from compromised identity servers.

We do not consider impractical credential storage and verification protocols that require users to maintain multiple credentials for identification, even if this is a naive solution to achieve survivable SSO.

We consider the following categories of authentication protocols that can be adopted to build a survivable SSO system:

- **P1:** a strawman approach based on plaintext storage and verification;
- **P2:** approaches that protect password storage but where verification is operated in plaintext [7];
- **P3:** approaches where passwords are protected at verification and storage time [18];
- **P4:** approaches that use secret sharing techniques to distribute passwords over multiple servers [3, 6, 26];

The presence of a majority of honest servers guarantees security against attacks B1 and B2 independently of the adopted authentication protocol and violation pattern. Even if the adversary registers malicious credentials (B1), he is unable to obtain enough valid partial tokens. Moreover, if the adversary prevents the completion of the authentication protocol in compromised identity servers (B2), the remaining honest majority ensures availability. If an authentication protocol is vulnerable to attacks B3 and B4, then the adversary can recover user credentials and impersonate him by compromising one identity server. Hence, violation patterns are not relevant to evaluate the security of protocols that are vulnerable to B3 and B4 because they already fail with one compromised sever.

P1. As a strawman approach, we consider a protocol in which each identity server stores and verifies plaintext passwords. The user sends the password over the secure channel to the identity server which verifies that it matches the stored password. This protocol yields a SSO protocol that is vulnerable to attacks B3, B4 and B5. The protocol is not survivable as an adversary that either compromises a single credentials database (B3) or observes the password verification procedure of one identity server (B4), can naively recover the password. The protocol is vulnerable to R5 (MITM) because, even if the communication channel is authenticated, the adversary may forward messages from compromised identity servers which are legitimate endpoints of the authenticated channel.

P2. Protocols that protect password storage but verify passwords in plaintext rely on password-hashing techniques [7]. Adopting these protocols in the considered distributed SSO scenario requires the user to transmit his password over a secure channel to each identity server which compares the password with its corresponding digest. This protocol is not completely secure against attack B3 because an adversary that captures the credentials database of any identity server can mount offline dictionary attacks (ODA). This may allow the attacker to recover the password to impersonate the user at the other identity servers. Moreover, the resulting SSO protocol is vulnerable to attacks B4 and B5 if an identity server is compromised. This protocol is vulnerable to B4 because a compromised identity server receiving the plaintext password can impersonate the user. It is also vulnerable to B5 because an adversary can forward the received plaintext password and impersonate the user to other identity servers.

P3. Security against server violation can be achieved through authentication protocols where identity servers never access plaintext passwords. The current state-of-the-art is represented by the OPAQUE scheme [18], which allows a user to store a secret key encrypted with his password at the registration phase. During authentication, only a user who knows the correct password can decrypt the secret key to prove his identity. The scheme adopts an oblivious PRF [14] which does not disclose any information about the password nor the secret key to the identity server during the registration and authentication phases. OPAQUE is not vulnerable to B5 because adopts channel bindings to prevent MITM attacks. While the SSO protocol obtained by different OPAQUE instances with each identity server is not vulnerable to B5, it is not completely secure against B3 and B4 because an adversary that captures the credentials database or observes the verification procedure of a single identity server may be able to mount offline dictionary attacks.

P4. We conclude by evaluating proposals that are secure against attacks B3, B4 and B5 relying on schemes based on secret sharing, such as Threshold Oblivious PRFs (TOPRF) [3,6,26]. The work of [3] is secure against these attacks considering static corruptions of identity servers, whereas [6,26] are secure in all violation patterns as they are proven secure against mobile adversaries. They are secure against B3 and B4 because they rely on techniques based on secret sharing to store and transmit the password. Hence, individual messages and credential databases do not contain enough information to mount attacks that can recover the password, such as offline dictionary attacks. These proposals are also secure against B5 because they are proven secure against active adversaries that can eavesdrop communications of up to a threshold of other identity servers.

Table 1. Security guarantees of survivable single sign-on systems with different authentication protocols

	B3 verification key	**B4** internal state	**B5** MITM
P1 - Strawman	○	○	○
P2 - Pwd hashing	◑	○	○
P3 - OPAQUE	◑	◑	●
P4 - Secret sharing	●	●	●

○: vulnerable, ◑: partially vulnerable (offline dictionary attack), ●: not vulnerable

The trade-offs of different authentication protocols are summarized in Table 1, where columns denote SSO requirements and rows denote the considered protocols. Password-based protocols that are not designed to be survivable, are either insecure against B3, or only partially secure. We note that when a protocol is partially secure against B3 due to possible offline dictionary attacks, the user can choose a strong password to make these attacks ineffective.

8 Conclusions

We propose the first flexible and survivable SSO protocol that relies on a distributed architecture of identity servers that collectively authenticate users and issue SSO tokens through a novel scheme. Flexibility allows service providers to choose the best trade-off between performance and security for each service and to preserve compatibility with non-survivable SSO. Survivability allows the identity provider to guarantee a high level of identity assurance even in presence of successful intrusions. We evaluate the security of the overall survivable SSO by considering several state of the art authentication protocols. Moreover, we show that the proposed token release scheme is secure against a comprehensive set of attack classes. Flexibility and survivability make the proposal a viable solution to offer secure and robust authentication to cloud services and any mission critical system that must rely on SSO. The results of this paper are open to different developments. It should be interesting to investigate how emerging passwordless authentication protocols may impact flexibility in the context of survivable SSO systems. Moreover, we think that it is possible to extend this proposal to support decentralized management systems as in the context of multi-cloud environments.

References

1. Regulation (EU) no 910/2014 of the European parliament and of the council of 23 July 2014 on electronic identification and trust services for electronic transactions in the internal market and repealing directive 1999/93/EC. Technical report, European Parliament, Council of the European Union (2014). https://eur-lex.europa.eu/eli/reg/2014/910/oj
2. Detecting abuse of authentication mechanisms. Technical report, National Security Agency (2020)
3. Agrawal, S., Miao, P., Mohassel, P., Mukherjee, P.: PASTA: PASsword-based threshold authentication. In: Proceedings of the ACM SIGSAC Conference on Computer and Communications Security (2018)
4. Andreolini, M., Pietri, M., Tosi, S., Balboni, A.: Monitoring large cloud-based systems. In: 4th International Conference on Cloud Computing and Services Science, CLOSER. SciTePress (2014)
5. Avizienis, A., Laprie, J., Randell, B., Landwehr, C.: Basic concepts and taxonomy of dependable and secure computing. IEEE Trans. Dependable Secure Comput. **1**(1), 11–33 (2004)
6. Baum, C., Frederiksen, T.K., Hesse, J., Lehmann, A., Yanai, A.: PESTO: proactively secure distributed single sign-on, or how to trust a hacked server. In: 5th IEEE European Symposium on Security and Privacy (2019)
7. Biryukov, A., Dinu, D., Khovratovich, D.: Argon2: new generation of memory-hard functions for password hashing and other applications. In: European Symposium on Security and Privacy. IEEE (2016)
8. Boeyen, S., Santesson, S., Polk, T., Housley, R., Farrell, S., Cooper, D.: Internet X.509 public key infrastructure certificate and certificate revocation list (CRL) profile. Technical report, IETF (2008)
9. Boldyreva, A.: Threshold signatures, multisignatures and blind signatures based on the gap-Diffie-Hellman-group signature scheme. In: Desmedt, Y.G. (ed.) PKC 2003. LNCS, vol. 2567, pp. 31–46. Springer, Heidelberg (2003). https://doi.org/10.1007/3-540-36288-6_3
10. Cash, D., Meltzer, M., Koessel, S., Adair, S., Lancaster, T.: Dark halo leverages solarwinds compromise to breach organizations. Technical report, Volexity (2020). https://www.volexity.com/blog/2020/12/14/dark-halo-leverages-solarwinds-compromise-to-breach-organizations/
11. Damgård, I., Jakobsen, T.P., Nielsen, J.B., Pagter, J.I.: Secure key management in the cloud. In: Stam, M. (ed.) IMACC 2013. LNCS, vol. 8308, pp. 270–289. Springer, Heidelberg (2013). https://doi.org/10.1007/978-3-642-45239-0_16
12. Eldefrawy, K., Ostrovsky, R., Park, S., Yung, M.: Proactive secure multiparty computation with a dishonest majority. In: Catalano, D., De Prisco, R. (eds.) SCN 2018. LNCS, vol. 11035, pp. 200–215. Springer, Cham (2018). https://doi.org/10.1007/978-3-319-98113-0_11
13. Fisher, W., et al.: Mobile application single sign-on: improving authentication for public safety first responders (2nd draft). Technical report, National Institute of Standards and Technology (2019). Special Publication 1800-13B
14. Freedman, M.J., Ishai, Y., Pinkas, B., Reingold, O.: Keyword search and oblivious pseudorandom functions. In: Kilian, J. (ed.) TCC 2005. LNCS, vol. 3378, pp. 303–324. Springer, Heidelberg (2005). https://doi.org/10.1007/978-3-540-30576-7_17
15. Garcia, M., Bessani, A., Gashi, I., Neves, N., Obelheiro, R.: Analysis of operating system diversity for intrusion tolerance. Softw. Pract. Exp. **44**(6), 735–770 (2014)

16. Grassi, P.A., Garcia, M.E., Fenton, J.L.: Digital identity guidelines. Technical report, National Institute of Standards and Technology (2017). DRAFT Special Publication 800-63c
17. Herzberg, A., Jarecki, S., Krawczyk, H., Yung, M.: Proactive secret sharing or: how to cope with perpetual leakage. In: Coppersmith, D. (ed.) CRYPTO 1995. LNCS, vol. 963, pp. 339–352. Springer, Heidelberg (1995). https://doi.org/10.1007/3-540-44750-4_27
18. Jarecki, S., Krawczyk, H., Xu, J.: OPAQUE: an asymmetric PAKE protocol secure against pre-computation attacks. In: Nielsen, J.B., Rijmen, V. (eds.) EURO-CRYPT 2018. LNCS, vol. 10822, pp. 456–486. Springer, Cham (2018). https://doi.org/10.1007/978-3-319-78372-7_15
19. Jhawar, R., Piuri, V.: Fault tolerance and resilience in cloud computing environments. In: Computer and Information Security Handbook. Elsevier (2017)
20. Magnanini, F., Ferretti, L., Colajanni, M.: Scalable, confidential and survivable software updates. IEEE Trans. Parallel Distrib. Syst. **33**(1), 176–191 (2022). https://doi.org/10.1109/TPDS.2021.3090330
21. Nikitin, K., et al.: CHAINIAC: proactive software-update transparency via collectively signed skipchains and verified builds. In: Proceedings of the 26th USENIX Security Symposium (2017)
22. Nystrom, M., Kaliski, B.: PKCS #10: Certification Request Syntax Specification Version 1.7. RFC 2986 (2000). https://rfc-editor.org/rfc/rfc2986.txt
23. Ostrovsky, R., Yung, M.: How to withstand mobile virus attacks. In: Proceedings of the Tenth Annual ACM Symposium on Principles of Distributed Computing (1991)
24. Ray, I., Belyaev, K., Strizhov, M., Mulamba, D., Rajaram, M.: Secure logging as a service-delegating log management to the cloud. IEEE Syst. J. **7**(2), 323–334 (2013)
25. Rose, S., Borchert, O., Mitchell, S., Connelly, S.: Zero trust architecture. Technical report, National Institute of Standards and Technology (2020)
26. Zhang, Y., Xu, C., Li, H., Yang, K., Cheng, N., Shen, X.S.: PROTECT: efficient password-based threshold single-sign-on authentication for mobile users against perpetual leakage. IEEE Trans. Mob. Comput. **20**(6), 2297–2312 (2021)
27. Zhou, L., Schneider, F.B., Van Renesse, R.: COCA: a secure distributed on-line certification authority. ACM Trans. Comput. Syst. (TOCS) **20**(4), 329–368 (2002)

The Analysis and Implication of Data Deduplication in Digital Forensics

Izabela Savić and Xiaodong Lin[(✉)]

University of Guelph, Guelph, Canada
{savici,xlin08}@uoguelph.ca

Abstract. Data deduplication is a file storage system method that is available on various operating systems such as Windows Server, MacOS, and Linux distributions. However, rehydration of deduplicated files is not yet a functionality supported in forensic tools. With the increasing cost of cybercrimes each year, and more users looking for ways to save storage space on ever growing file sizes, developing forensic tools to support detection and recovery of deduplicated files is more important than ever. To address this issue, in this paper, we first give a comprehensive analysis of data deduplication and its implementation in modern Operating Systems. Then, we examine how data deduplication techniques affect digital forensic investigation tools, particularly, TSK, a widely used open-source forensic tool for volume and file system analysis. We then propose a solution to restore deduplicated files from an acquired deduplicated file system volume and implement it into TSK to extend TSK for supporting data deduplication. Furthermore, we also study the positive forensic implications of data deduplication techniques, which are seldom considered in existing studies. Specially, we also investigate new sources of evidence or new artifacts generated during data deduplication.

Keywords: Data deduplication · Digital forensics · TSK · Windows server 2012 · OpenDedup

1 Introduction

Cybercrimes are an extremely common and prominent form of computer abuse in today's society. It is estimated that cybercrime will cost the world roughly $6 trillion annually by the year 2021, this is a $3 trillion dollar increase from 2015. Theft of intellectual property, personal and financial data, embezzlement, forensic investigation, and restoration of deleted and hacked systems are all factors which are included in the $6 trillion estimate. It has even been concluded that cyber-attacks are now the fastest growing crime in the United States [1]. Many of these cyber-attacks have made the news, such as the Heartbleed vulnerability which led to approximately 900 social insurance numbers being stolen from the CRA [2], or the more recent SolarWinds attack which allowed attackers to gain access to thousands of devices. In the event of cyber-attacks, defense is not always the answer, which is where computer forensics comes into play.

© Springer Nature Switzerland AG 2022
W. Meng and M. Conti (Eds.): CSS 2021, LNCS 13172, pp. 198–215, 2022.
https://doi.org/10.1007/978-3-030-94029-4_14

Computer forensics, which is described as "the science of identifying, extracting, and preserving computer logs, files, cookies, cache, meta-data, internet searches, and any other legally admissible evidence that could be used to solve crimes committed using internet connected infrastructure" [3], can be used to cultivate evidence of a crime from a machine aiding in the apprehension and incarceration of a criminal. This is partially possible due to the file system, which tracks a multitude of events on the system leaving a forensic footprint for an investigator to analyze. Activities such as accessing or changing a file, creating or deleting a file, increase or decrease in space utilization, memory that is written to a page file, and a log of process events create an entry in the file system log [3]. Analysis of the file system is critical to an investigator, as it provides them with a "wealth of information about what happened to the system" [3], allowing investigators to reconstruct the events which happened on the machine.

The investigation of the file system is typically the first step in a computer forensic investigation, as the file system will "yield more information about what happened than any other sources available" [3]. Acquiring the file system, while maintaining its original state, is of the utmost importance to a computer forensics investigation, as the file system can hold "footprints" of network intrusions, malware installations, file deletions, and more. For example, through analyzing the file system, we can recover deleted files. When a file is deleted, typically the file descriptor is deleted, however, the bits of data on the disk are not overwritten right away [4]. The computer recognizes the spaces in which a deleted file resided as "free" and overwrites the bytes with time as the user continues to update and install/add more files. By analyzing the file descriptor, for example, in the File Allocation Table (FAT) file system, we can retrieve the start position and size of the file and retrieve the corresponding bytes from the disk, as the majority of files are stored contiguously. So long as they have not been overwritten, we've now very easily retrieved a deleted file [5].

Over the last few decades, operating systems and file system management have greatly changed. As the size of files continues to increase, so does the demand for larger storage and technologies which make more efficient use of disk space. Technologies such as file compression and data deduplication allow users to retain a more efficient use of their disk space but come at the cost of introducing new challenges for file system forensics and data recovery. For example, data deduplication, an emerging technology allowing users to no longer store redundant copies of data to decrease disk space usage, is providing computer forensic investigators with a unique challenge when it comes to file system forensics. Rehydration, the process of reversing the data deduplication, appears to some as a promising possibility, but poses more harm than good and is a discouraged approach. Rehydrating a volume could possibly overwrite unallocated or deleted data and is not always an available option if the unoptimized volume is too large to fit on the current disk. Rehydrating also carries an unnecessary risk of possibly introducing a system failure [6]. When conducting AV (antivirus) scans and attempting to analyze a disk image using carve data or key word search, one will encounter "errors occurred relating to file being inaccessible" [7]. As of

2017, none of the major forensic suites handled data deduplication, as of now there still is no evidence of any forensic suite handling data deduplication [7].

The goal of our research is to examine the effect of data deduplication techniques, particularly their effects on opensource digital forensic investigation tools used for volume and file analysis (e.g., TSK), and data recovery roles (e.g., ProDiscover). Through our research, we aim to develop a solution that would allow an application such as TSK to support and remain functional when encountering volumes with data deduplication enabled. In particular, the contributions of this work are as follows.

- We examine the effect of data deduplication techniques on open source digital forensic investigation tools. Then, we design and develop a data deduplication rehydration algorithm to extend TSK, a widely used open source file system forensic tool, for supporting processing deduplicated files.
- We explore the positive implications of data deduplication from a forensic standpoint. Due to the nature of data deduplication, and how more than one file can reference the same chunk, we have explored whether deduplication can allow the restoration of chunks of multiple files, rather than only the original file targeted. Extensive experiments show that in some circumstance, it could allow forensic analysts to recover more artifacts than originally believe possible.

The remainder of this paper is organized as follows. We will begin with a more in-depth detail of data deduplication in Sect. 2, followed by a summary and analysis of TSK tools and functions in Sect. 3. Section 3 will also include a detailed explanation of the testing setup. In addition, we will also include our proposed solutions in Sect. 3, as we aim to make our solution for forensic software compatible with widely used data deduplication tools, if not compatible with all data deduplication tools. We will explore the positive implications of data deduplication in Sect. 4. Then, we discuss some related works in Sect. 5. Finally, the conclusion and final remarks on the future work are given in Sect. 6.

2   Data Deduplication: An in Depth Examination

2.1   Data Deduplication - An Overview

Data deduplication is a process that decreases the storage capacity requirements by eliminating multiple copies of data [8]. This is done by replacing duplicate data blocks with a pointer that points to the physical location of the first copy of that block [9]. Before we proceed further, we will present a list of data deduplication terminology and definitions, shown in Table 1.

After discussing the necessary terminology, we can now proceed to discuss the categorizing of data deduplication techniques. Data deduplication techniques can be divided into 2 categories: inline deduplication and post-processing deduplication, from there on they can be further divided into 3 categories: file-level,

Table 1. Data deduplication terminology.

Term	Definition
Cluster	A group of consecutive sectors, aka block or data unit. File and directory content are stored in clusters, as clusters are the smallest disk space allocated to files in a file system. Note that TSK uses cluster addresses to examine data except in FAT where sector addresses are used instead
Chunk	A consecutive number of N KB, where N can be variable or fixed length. Typically contain one or more contiguous clusters. Each individual chunk is referenced by a hash pointer where the hash is calculated on the data contained within it
Chunk store	Directory/Segment of memory allocated to storing of chunks
Hash	A hashing algorithm will take input and try and make a unique value for each input. Basically, taking arbitrary length data and performing a calculation on it and the hash is the output, which has a fixed-length
Hash-db	It contains the hashes of all the chunks in the chunk store
Hash sequence	Hash sequences are incredibly important in data deduplication. A hash sequence is a sequence of hashes produced by hashing each chunk. File are replaced with hash sequences that map to the corresponding chunk in chunk store
Hotspot container	Stores most common/frequently used chunks. This feature was added by Microsoft for Windows Server 2012 & 2016
Rehydrating	Process of taking deduplicated file which is a file full of hashes and reconstructing the file with the proper memory contents
Reparse point	Acts as a collection of user-defined data

block-level, and byte-level. Inline deduplication is a method of doing data deduplication before the file is written to the disk. The data is temporarily written to a buffer where the file is then split into chunks of a given size, and a hash is calculated for each chunk [10]. If a chunk with the same hash does not already exist on the system, the chunk is written to disk, and the hash-db is updated to contain the new hash [10]. The file is also updated to include the hash [10]. If the chunk's hash already exists on the disk, the hash is added to the hash sequence of the file, the map is updated to include a pointer connecting the file and physical location of the chunk in memory, and the chunk is freed from the buffer [10].

In post-processing deduplication, the data is written to the disk immediately as it would be in a regular file system. After a set period of time, the original files are processed, where they will be split into chunks, have their hashes computed, and are checked against the hash-db as mentioned above [9]. Then redundant pieces of the file are removed from the disk. Data deduplication at a file-level works by analyzing files or "structured data" (e.g. emails, databases, etc.) and removing duplicates. They are particularly useful for email repositories (can remove duplicate emails), databases, and virtual machine image backups [11]. Block-level deduplication allows us to analyze blocks of a file and remove duplicates [11]. Blocks can be viewed as partitions of a larger data or sections of a file. Lastly, byte-level deduplication is a form of block-level deduplication that is "content aware", meaning that some reverse engineering on the data stream has been performed to retrieve bytes corresponding to information used, such as the file name, file type, etc. [12].

Both inline and post-processing deduplication methods have their strengths and weaknesses. Inline deduplication requires less disk space but can create a bottleneck in the deduplication processing, whereas post-processing deduplication requires more disk space while the data is being processed and duplicate data has not been freed yet but does not have a bottleneck due to processing [9].

Regardless of whether using inline or post-processing deduplication, both follow the same methodology. A file is split into chunks, a hash is computed for each chunk, if the hash exists in hash-db the file hash sequence is updated the chunk is freed, if it does not, the file hash sequence is updated, and the chunk is stored. Please note that in certain deduplication systems files with certain properties, such as compressed files, are not deduplicated but stored as they would be in a normal file system. A general visual representation of data deduplication systems can be seen in Fig. 1.

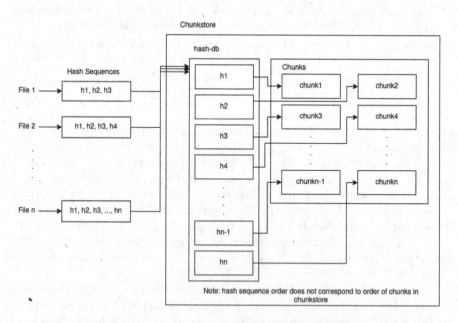

Fig. 1. Data deduplication general diagram.

Overall, data deduplication helps decrease storage usage but comes at the cost of added overhead and the possibility of decreased performance [9]. It also comes with the risk that if a block of data becomes corrupted, all files pointing to that block will have lost the original data.

2.2   OpenDedup vs. Windows Server 2012

There are various data deduplication systems currently available. However, we will be focusing on OpenDedup, and Windows Server 2012 and 2016's deduplication system.

OpenDedup. OpenDedup is an example of an inline data deduplication system. It is compatible with Ubuntu 14.01+ and various other operating systems [13]. All necessary files in the regular file system are allocated using FUSE (Filesystem in USErspace), the SDFS, a deduplicated file system, appears like a mounted drive on Linux, and files are accessed through the mount point [10]. For OpenDedup, the chunk size is a fixed size, and the hash function used is Murmurhash3 with a 128-bit length and seed 0x192A [14]. Inside the pointer file, the first bytes contain information regarding the creator, the last bytes contain the version of the file. Inside the file is the size of the original file, along with a unique identifier pointing to elements in the structure (folders) that contain the sequence of chunks that make up the file [14]. Each folder is named with a unique identifier and has its own map file, which acts as a hash-db, containing all the hashes of all the chunks stored in that folder. Folders are named in the following format: "nnn-nnn" (e.g., 123–125) where chunks in the folder have a hash that fits in that range [14]. The map files follow a structure which is the key, length (length of chunk size + length of start marker) and value (the chunk) [14]. In OpenDedup, a folder acts as a chunk store. To further understand how OpenDedup handles data deduplication, we can take a look at Fig. 2.

Windows Server 2012 and 2016. The data deduplication systems found on Windows Server 2012 and 2016 are an example of post-processing deduplication. The files are stored in the regular file system, and after a set period the files are deduplicated using SHA256 for hashing, and the original file is removed [14].

In Windows Server 2012 & 2016 not all files are deduplicated. Files smaller than 32 KB, system state files, encrypted and/or compressed files, and files with certain extensions are saved regularly and never deduplicated [14].

When deduplicating, depending on the system configuration, chunks are compressed when stored. However, this does not apply to already compressed file chunks [14].

All deduplicated data is stored in the system volume information folder, and each chunk store has 3 folders/containers: stream container, data container, and hotspot container. The Windows file system or NTFS (New Technology File System) contains a master file table ($MFT), where each entry related to a deduplicated file holds information about the chunkstore which is saved as

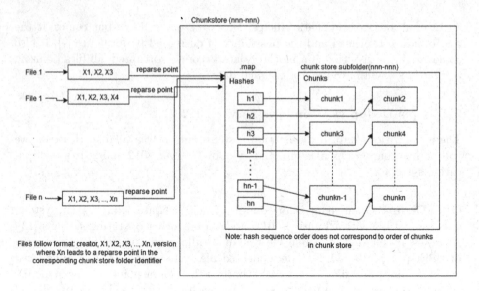

Fig. 2. Opendedup file deduplication process.

a "reparse point" ($REPARSE_POINT - 0xC0) [14]. The reparse point starts with NTFS attribute type 0xC0, the length of the file is written at byte offset 0x28, the chunk store identifier is at byte offset 0x38, and the stream header is at byte offset 0x78 [14]. The stream header is used to identify the correct hash entry in the stream file. A chunk store folder contains a data folder and a stream folder [14]. The data folder is where chunks are stored and contains a .ccc file which is the chunk container [14]. The stream folder contains the hash sequences for the deduplicated files and has a .ccc file with 3 sections: cthr - file header, rrtl - redirection folder, and ckhr - streammap element. The chkr section contains the full sequence of hashes relative to the file (a stream map) [14]. Similarly, to Fig. 2, we can view a file deduplication process diagram for Windows Server 2012 in Fig. 3. We can also view a sample MFT table entry in Fig. 3. It is also important to note that a hash entry can contain more than one sequence, each sequence has a size of 0x60 bytes, starting at byte offset 0x70 bytes from the beginning of the hash entry. The sequences are stored sequentially as shown by an increasing sequence number across all sequences (0x01, 0x02, ... 0xFA, ... etc.) [14].

We can also note that data deduplication is not supported on system or boot volumes, remote mapped or remote mounted drivers, cluster shared volume system (CSVFS) for non-VDI workloads or any workloads on Windows Server 2012, files approaching or greater than 1 TB, and volumes approaching or larger than 64 TB in size [15].

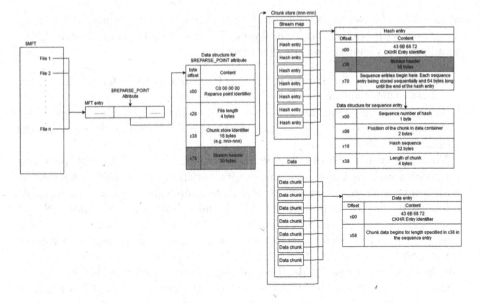

Fig. 3. Windows server 2012 deduplication overview.

3 Affecting File System Forensic Analysis Effectiveness

In this section, we will examine how data deduplication affects existing file system forensic tools. There are a multitude of computer forensic tools available to investigators, such as TSK (The Sleuth Kit) [16], The Forensic Toolkit [17], Toolsley [18], and many more. The tools provided by TSK are used by various individuals around the world, along with companies, such as Cyber Triage [19].

TSK provides multiple tools to users such as deleted file retrieval, scanning for malicious files, key word file search, as well as file content viewing. It provides a wide selection of data source support and accepts data sources such as disk images, VM files, local disks, local files/folders, and unallocated space images [16].

Due to the wide variety of usage of TSK, the fact that it is a free open source tool, and supports multiple data source types, we will be analyzing the effects of data deduplication from a disk image on TSK tools. We will first begin by discussing the various TSK tools available, their functionality, and how TSK tools are categorized. Next, we will discuss the performance of TSK tools when tested against retrieving file information from a deduplicated file. Afterward, we will present an enhancement to TSK that enables the analysis of data deduplication enabled file system.

3.1 TSK Tools

The tools provided in TSK for manual analysis can be split into six sections: image file tools, volume system tools, file system layer tools, file name layer tools, metadata layer tools, and data unit layer tools. Each of these tools is used to interact with one of four TSK layers: file system layer, content/data layer, metadata layer, human interface/file layer.

- **Image File Tools.** The image file tools "contains tools for the image file format" [20]. The image file tools division contains tools such as img_stat, and img_cat. img_stat is used to show the details of the image format using the appropriate command line flags, and img_cat will display the raw contents of an image file.
- **Volume System Tools.** The volume systems tools "take a disk (or other media) image as input and analyze its partition structures". The volume system tools division contains tools such as mmls, mmstat, and mmcat. mmls is used to display the layout of the disk and will display both allocated and unallocated spaces. mmstat, similarly to img_stat, will display the details about a volume system. Lastly, mmcat will display the contents of a volume using STDOUT [20].
- **File System Layer Tools.** The file system layer tools "process general file system data, such as the layout, allocation structures, and boot blocks" [20]. The file system layer tools division contains tools such as fsstat, which is used to show file system details and statistics. This includes layout, sizes, and labels.
- **File Name Layer Tools.** The file name layer tools "process the file name structures, which are typically located in the parent directory." [20]. The file name layer tools division contains tools such as ffind and fls. ffind is used to find allocated and unallocated file names that point to a given metadata structure, and fls lists all allocated and unallocated file names in a directory [20].
- **Metadata Layer Tools.** The metadata layer tools "process the meta data structures, which store the details about a file" [20]. The metadata layer tools division contains tools such as icat and ifind are used to process the metadata structures. icat is used to extract the data units of a file and is specified using the metadata address instead of the file name [20]. ifind is used to find the metadata structure according to a given file name or data unit [20].
- **Data Unit Layer Tools.** Lastly, the data unit layer tools "process the data units (clusters or blocks) where file content is stored" [20]. The data unit layer tools division contains tools such as blkcat, blkls, blkstat, blkcalc, and are used to process data units of the file [20]. blkcat will extract the contents of a given data unit, blklks will list details about data units, blkstat will display statistics about the data units, and blkcalc will calculate where data in the unallocated space image exists [20].

3.2 Effects of Data Deduplication on File System Forensics

Next, we examine how data deduplication affects TSK tools. Using VMWare [23], a Windows Server 2012 VM was initialized. An additional 20 GB partition with data deduplication enabled was included in Windows Server. The deduplicated volume had NoCompress set to true, and the deduplication process was scheduled for 11:00 pm each evening. The VM image was stored as one image. 5 files were downloaded and placed onto the deduplication enabled partition: 1342-0 (.txt) [24], sample-mp4-file (.mp4) [25], get_started_with_smallpdf (.pdf) [26], Symphony No. 6 (1st movement) (.mp3) [27], and Sample_1280x720_surfing_with_audio (.mkv) [28]. After the scheduled deduplication had run, the VM image was then converted to an .img file (from a .vmdk) using qemu-img [22], and then processed using Autopsy.

Due to the nature of TSK tools, it is our prediction that all tools which require figuring out address of a data unit (cluster) allocated to a deduplicated file will fail, due to following the incorrect data address. To support our assumptions, we tested TSK tools on our image with data deduplication enabled, and the results can be seen in Table 2. For example, istat, which uses the file's inode number and displays its details including the addresses of data units occupied by the file, will fail. It can be observed in Fig. 4 that all addresses (contained in the $DATA attribute for a regular file) are zero.

Table 2. TSK tool analysis.

Tool layer	Tool name	Unaffected by data deduplication?
File system layer tools	fsstat	Yes
File name layer tools	ffind	Yes
	fls	Yes
	fcat	**No**
Meta data layer tools	icat	**No**
	ifind	Yes
	ils	Yes
	istat	**No**
Data unit layer tools	blkcat	Yes
	blkls	Yes
	blkstat	Yes
	blkcatc	Yes
Volume system tools	mmls	Yes
	mmstat	Yes
	mmcat	Yes
Image file tools	img_stat	Yes
	img_cat	Yes

```
MFT Entry Header Values:
Entry: 48        Sequence: 2
$LogFile Sequence Number: 33649933
Allocated File
Links: 1

$STANDARD_INFORMATION Attribute Values:
Flags: Archive, Sparse, Reparse Point
Owner ID: 0
Security ID: 264   (S-1-5-32-544)
Created:        2021-08-15 23:17:11.657738100 (Eastern Summer Time)
File Modified:  2021-08-09 20:23:50.000000000 (Eastern Summer Time)
MFT Modified:   2021-08-15 23:17:11.657738100 (Eastern Summer Time)
Accessed:       2021-08-15 23:17:11.657738100 (Eastern Summer Time)

$FILE_NAME Attribute Values:
Flags: Archive
Name: 1342-0.txt
Parent MFT Entry: 5      Sequence: 5
Allocated Size: 802816         Actual Size: 0
Created:        2021-08-15 23:17:11.657738100 (Eastern Summer Time)
File Modified:  2021-08-15 23:17:11.657738100 (Eastern Summer Time)
MFT Modified:   2021-08-15 23:17:11.657738100 (Eastern Summer Time)    Correctly
Accessed:       2021-08-15 23:17:11.657738100 (Eastern Summer Time)    listed file size

Attributes:
Type: $STANDARD_INFORMATION (16-0)   Name: N/A   Resident   size: 72
Type: $FILE_NAME (48-2)   Name: N/A   Resident   size: 86
Type: $DATA (128-1)   Name: N/A   Non-Resident, Sparse   size: 799645  init_size: 799645
0 0 0 0 0 0 0 0
0 0 0 0 0 0 0 0
0 0 0 0 0 0 0 0            Incorrect all zero data unit (cluster)
. . . . . .               addresses. 208 total clusters.
0 0 0 0 0 0 0 0
0 0 0 0 0 0 0 0
0 0 0 0 0 0 0 0
Type: $REPARSE_POINT (192-3)   Name: N/A   Resident   size: 132
```

Fig. 4. istat output on a deduplicated file [21].

3.3   Extending TSK

In this section, we will begin by introducing the implementation of a selected
TSK tool and discussing our implemented solution which enables the tool to function properly as intended on a deduplicated file. The tool we've selected to analyze was istat. istat works by following the MFT table entry and parsing the
attributes included in the entry. For example, we can see in Fig. 4 that the included
attributes are the $STANDARD_INFORMATION, $FILE_NAME, $DATA, and
$REPARSE_POINT attributes. With the current functionality of istat, istat will
follow the $DATA information regardless of if the $REPARSE_POINT is present.
This is incorrect. If a $REPARSE_POINT attribute is present then the file is dedu-
plicated [29], thus rendering the information of the $DATA attribute useless. To
correct this, we propose to write a module that will check if the file entry has a
$REPARSE_POINT attribute. If it does, it will not print the $DATA attribute
contents, and instead will follow the chunkstore identifier to the correct folder and

find the hash entry in the stream map. Using the hash entry, we will then have the position of the chunk in the data container and can retrieve the chunk to rehydrate the file. We can see a visual representation of this algorithm in Fig. 5.

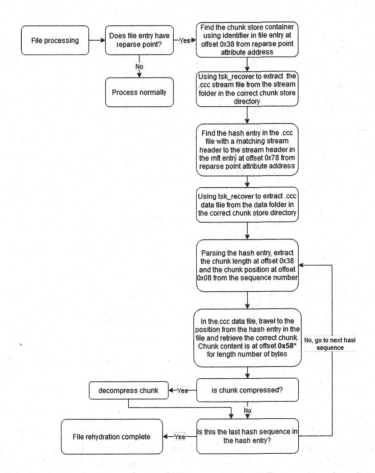

Fig. 5. Proposed rehydration process. *Please note the difference in values here as the chunk content starts at byte offset 0x58 as opposed to 0x5C in [14].

Through the use of our rehydration algorithm presented in Fig. 5, we were able to rehydrate the 1342-0.txt file. Figure 6 is a screenshot of a part of a hash entry in the stream map file. At file offset 0x5070 we have the first sequence entry of the hash. The length of the chunk associated with the sequence entry is 4 bytes located at 0x50a8 (or 0x38 bytes from the start of the sequence entry at 0x5070), and the absolute location of the chunk in the chunk container is 4 bytes started at 0x5078 (or 0x08 bytes from the start of the sequence entry at 0x5070). This pattern is repeated for each sequence entry (e.g., 0x50b0, 0x50f0).

If we use the absolute location at 0x5078 as the file offset in the data container .ccc file, which can be seen in Fig. 7, we will find our first chunk. We can note that the chunk content begins at byte offset 0x58 from the start of the data entry (0x5000).

```
0x00005070: 01 00 00 00  01 00 00 00  00 50 00 00  01 00 00 00   .........P......
0x00005080: AB B6 00 00  00 00 00 00  AC 06 49 D1  92 5F 34 A9   ..........I.._4.
0x00005090: 69 F4 2B D9  5A 2C C0 24  26 88 96 9F  2D 0F 2F 7B   i.+.Z,.$&...-./{
0x000050a0: B4 40 F1 35  A1 BB BE F1  AB B6 00 00  00 00 00 00   .@.5............
0x000050b0: 02 00 00 00  01 00 00 00  08 07 01 00  01 00 00 00   ................
0x000050c0: 58 B1 01 00  00 00 00 00  4A 6C EE 33  4D CE D7 04   X.......J1.3M...
0x000050d0: 35 3E 8C CD  A0 42 9E A2  2B AD 00 4E  E4 FA 00 A7   5>...B..+..N....
0x000050e0: 8A D8 0A 3D  2D D5 E9 4F  AD FA 00 00  00 00 00 00   ...=-..O........
0x000050f0: 03 00 00 00  01 00 00 00  10 02 02 00  01 00 00 00   ................
0x00005100: 81 21 03 00  00 00 00 00  9B 1C 32 28  E3 3A A6 2A   .!........2(.:.*
0x00005110: 84 06 87 D0  E7 14 47 E4  4A DC 3C 52  CF 54 C7 49   ......G.J.<R.T.I
0x00005120: 7A 5B 65 FD  CB 74 C8 70  29 70 01 00  00 00 00 00   z[e..t.p)p......
```

Fig. 6. Stream map hash entry.

```
0x00005000: 43 6B 68 72  01 03 03 01  01 00 00 00  AB B6 00 00   Ckhr............
0x00005010: 01 00 28 00  08 00 00 00  08 00 00 00  08 00 00 00   ..(.............
0x00005020: 00 00 00 00  00 00 00 00  AC 06 49 D1  92 5F 34 A9   ..........I.._4.
0x00005030: 69 F4 2B D9  5A 2C C0 24  26 88 96 9F  2D 0F 2F 7B   i.+.Z,.$&...-./{
0x00005040: B4 40 F1 35  A1 BB BE F1  D0 A7 62 BB  34 B7 4E E9   .@.5......b.4.N.
0x00005050: 40 67 B0 2A  B9 B3 8A AE  EF BB BF 54  68 65 20 50   @g.*.......The P
0x00005060: 72 6F 6A 65  63 74 20 47  75 74 65 6E  62 65 72 67   roject Gutenberg
0x00005070: 20 65 42 6F  6F 6B 20 6F  66 20 50 72  69 64 65 20    eBook of Pride
0x00005080: 61 6E 64 20  50 72 65 6A  75 64 69 63  65 2C 20 62   and Prejudice, b
```

Fig. 7. Data container data chunk.

Through our efforts to improve TSK, a new sample output can be provided in Fig. 8, which is used to replace all zero data unit addresses in Fig. 4. This output utilizes the data units of the data chunk container file provided by istat and includes the starting and ending byte of a chunk in the data unit. For example, assume that the size of a data unit (cluster size) is 4 KB, and if a chunk starts at byte offset 55 after the beginning of a data unit and lasts until the end of the data unit, the data unit output will be data_unit:55-4095. Similarly, if a chunk occupies a data unit by starting at the beginning of a data unit and ends at byte offset 55 the data unit output will be data_unit:0-55. A single data unit address indicates that the entire data unit is occupied by a chunk of the file.

```
Chunk format: data_unit:data_unit_start_location-data_unit_end_location
```

779572:0088-4095	779573	779574	779575	779576
779577	779578	779579	779580	779581
779582	779583:0000-1794	779583:1888-4095	779584	779585
779586	779587	779588	779589	779590
779591	779592	779593	779594	779595
779596	779597	779598	779599:0000-0524	779599:0616-4095
779600	779601	779602	779603	779604
779605	779606	779607	779608	779609
779610	779611	779612	779613	779614
779615	779616	779617	779618	779619
779620	779621	779622:0000-0656	779622:0752-4095	779623
779624	779625	779626	779627	779628
779629	779630	779631	779632	779633
779634	779635	779636	779637	779638
779639	779640	779641	779642	779643
779644	779645	779646	779647	779648
779649	779650	779651:0000-3505	779651:3600-4095	779652
779653	779654	779655	779656	779657
779658	779659	779660	779661	779662
779663	779664	779665	779666	779667
779668	779669	779670	779671	779672
779673	779674	779675	779676	779677
779678	779679	779680	779681:0000-3639	779681:3728-4095
779682	779683	779684	779685	779686
779687	779688	779689	779690	779691
779692	779693	779694	779695	779696
779697:0000-3672	779697:3768-4095	779698	779699	779700
779701	779702	779703	779704	779705
779706	779707	779708	779709	779710
779711	779712	779713	779714	779715
779716:0000-2522	779716:2616-4095	779717	779718	779719
779720	779721	779722	779723	779724
779725	779726	779727	779728	779729
779730	779731	779732:0000-2929	779732:3024-4095	779733
779734	779735	779736	779737	779738
779739	779740	779741	779742	779743
779744	779745	779746:0000-4059	779747:0056-4095	779748
779749	779750	779751	779752	779753
779754	779755	779756	779757	779758
779759	779760	779761	779762	779763
779764	779765	779766	779767:0000-1847	

Fig. 8. TSK istat deduplicate file analysis improved output. *Please note if there is no data unit start location and data unit end location, the entire data unit is occupied by the file.

4 Positive Implications of Data Deduplication

Regardless of researcher's and worker's opinion or fears of data deduplication, it is a new method of file storage that will be here to stay. It is important that we look to utilize the possible positive implications of data deduplication, as this new storage method may benefit us more than we'd expect it to. Due to the nature of data deduplication, and how only one chunk is stored with each file containing a pointer to the single chunk, this means that we may now be able to recover more data than originally predicted. This could bring dozens of new forensic findings to light that possibly would not have been found in a forensic analysis of a non deduplicated file system.

Depending on the deduplication command run, we may even be able to assemble old versions of the file. When changing the 1342-0.txt file, old chunks were still in the data chunk file and could be retrieved. To further test and explore the positive implications of data deduplication we decided to test what happens to chunks after deleting a file, as well as editing a file. To conduct these tests, we took our original image and made two copies. In the first copy, the 1342-0.txt file was deleted, and a replica of the file was downloaded and placed on the deduplicated volume. An optimization followed by a garbage collection job was run on the deduplicated volume. For the second test, the 1342-0.txt file of the original image was modified (the first 21 chapters of the book were removed), and an optimization and garbage collection job was run on the volume. Before conducting each of these tests we made sure to note the stream container of the 1342-0.txt followed by the address of the stored chunks in the new stream container.

When observing the results of the first test, we can note that the stream container of the hash entry relating to the 1342-0.txt has changed. When comparing the new hash entry to the deleted file hash entry, we can see that the offset of the chunks in the data container are also different. This means that even with the file being deleted, and an exact replica of the file being put onto the deduplicated volume, the new file is being stored in a separate place. However, when going to the location of the old chunks we can now see that they've been overwritten by another file's data chunk, which will make recovering older entries harder in this case. When observing the results of the second test, we can once again note that the stream container of the hash entry relating to the 1342-0.txt has changed. We can also once again note that the offset of the chunk in the data container are also different. Unlike the first test, when visiting the old chunk locations of the original file versions the deleted chunks could be retrieved. In this case, we were fortunate enough that the chunks had not been overwritten by the chunk of another file.

Overall, we can conclude that under the right circumstances, it is possible to retrieve older chunks of deleted or edited files even very easily after a garbage collection deduplication job has been run on the volume. In the event that we are not fortunate enough for old chunks to not be overwritten further investigation and additional manipulation may be required.

This is truly an exceptionally positive implication of data deduplication that should be further investigated. Data deduplication could allow for additional artifacts to be produced from the same volume. Since multiple files may share multiple chunks, an investigator could be able to restore additional portions if not entire files that were not currently being investigated. In addition, we may also be able to restore older versions of important files. For example, if a file was deleted using a secure method deletion, there may be hope for a forensic investigator to retrieve older chunks corresponding to the file that have not been overwritten. Such additional findings could provide incredibly important information and evidence to investigators further allowing the perpetrator of a crime to be apprehended.

5   Related Works

In this section, we will be discussing related works which were deemed to be important regarding the research presented. The research presented by Lanterna and Barili in [14] was an integral building block of the data deduplication within this work. [14] was one of the few research papers in the field published which explained the format of data deduplication in Windows Server 2012 with precise detail. The figures which visually represented the hex-dump of the stream container and data containers, along with the note of offsets for the correct attributes (e.g., stream container, chunk length, chunk address, etc.) enabled us to further understand the process of data deduplication in Windows Server 2012, along with the process of how and where chunks were stored. [14] provided our research with a strong base, allowing us to develop our algorithm to rehydrate deduplicated files.

6   Conclusion

In this last section, we will discuss methods in which our research can be expanded to further improve forensic support for data deduplication techniques. Due to the constant need for more disk space, and the continuous size growth of files, it is safe to conclude that data deduplication will continue to become a more popular method of storing data in the future. Although this revelation may seem grim at first, primarily due to the lack of data deduplication support in most forensic tools, there may also be a positive note to this fact.

However, with the algorithm presented in Sect. 3, a forensic worker should be able to rehydrate a file using the MFT entry of the file, along with the stream file and data file from the correct chunk store folder. In addition, with the research findings presented in Sect. 4, we can further include that there is a possibility of workers being able to retrieve old chunks from deleted and updated files using the data file from the chunk container. This positive implication on data deduplication could potentially allow a forensic analyst to recover dozens of more artifacts for analysis possibly leading to large breakthroughs in forensic cases.

Nevertheless, a continuation of research into data deduplication and development of rehydration tools for modern forensic tools is extremely important and should be further researched and developed upon. Data deduplication is a file storage system that is here to stay and ignoring its existence would be incredibly irresponsible of forensic workers and researchers.

Acknowledgments. This work was supported by Natural Sciences and Engineering Research Council of Canada (NSERC).

References

1. Official Annual Cybercrime Report. Herjavec Group. https://www.herjavecgroup.com/wp-content/uploads/2018/12/CV-HG-2019-Official-Annual-Cybercrime-Report.pdf. Accessed 17 Feb 2021
2. Heartbleed. Wikipedia. https://en.wikipedia.org/wiki/Heartbleed. Accessed 17 Feb 2021
3. Brager, K.: Forensic Sources Part 3: File Systems. https://www.keirstenbrager.tech/file-systems. Accessed 17 Feb 2021

4. Lin, X.: Introductory Computer Forensics: A Hands-on Practical Approach. Springer, Cham (2018). https://doi.org/10.1007/978-3-030-00581-8
5. Changes for data Recovery. UFS Explorer. https://www.ufsexplorer.com/articles/chances-for-data-recovery.php. Accessed 17 Feb 2021
6. Harrison, A.: Forensic Analysis of Volumes with Data Deduplication Enabled. https://blog.1234n6.com/2017/08/forensic-analysis-of-volumes-with-data_29.html. Accessed 17 Feb 2021
7. Harrison, A.: Windows Server Data Deduplication and Forensic Analysis. https://blog.1234n6.com/2017/08/windows-server-data-depupliction-and.html. Accessed 17 Feb 2021
8. What is Data Deduplication? NetApp. https://www.netapp.com/data-management/what-is-data-deduplication/. Accessed 17 Feb 2021
9. Adshead, A.: Inline deduplication or post-process: which one is right for your environment? ComputerWeekly. https://www.computerweekly.com/tip/Inline-deduplication-or-post-process-Which-one-is-right-for-your-environment. Accessed 17 Feb 2021
10. Documentation. OpenDedup. https://opendedup.org/odd/documentation/. Accessed 17 Feb 2021
11. Campbell, M.: Byte-Level Versus Block-Level Deduplication and Backup. Unitrend. https://www.unitrends.com/blog/byte-level-versus-block-level-deduplication-backup. Accessed 17 Feb 2021
12. Whitehouse, L.: Data deduplication methods: block-level versus byte-level dedup. https://searchdatabackup.techtarget.com/tip/Data-deduplication-methods-Block-level-versus-byte-level-dedupe. Accessed 17 Feb 2021
13. Linux Quickstart Guide. OpenDedup. http://opendedup.org/odd/linux-quickstart-guide/. Accessed 17 Feb 2021
14. Lanterna, D., Barili, A.: Forensic analysis of deduplicated file systems. Digit. Invest. **20**(supplement), 99–106 (2017)
15. What's New in Data Deduplication. Microsoft. https://docs.microsoft.com/en-us/windows-server/storage/data-deduplication/whats-new. Accessed 03 Mar 2021
16. Sleuthkit. https://www.sleuthkit.org/. Accessed 03 Mar 2021
17. Exterro. https://www.exterro.com/forensic-toolkit. Accessed 03 Mar 2021
18. Toolsey. https://www.toolsley.com/. Accessed 03 Mar 2021
19. The Sleuth Kit. Cyber Triage. https://www.cybertriage.com/integration/the-sleuth-kit/. Accessed 14 Apr 2021
20. Gurram, V.: Introduction to The Sleuth Kit (TSK). Self-Published (2016)
21. istat. Sleuthkit. http://www.sleuthkit.org/sleuthkit/man/istat.html. Accessed 14 Apr 2021
22. qemu-img. https://linux.die.net/man/1/qemu-img. Accessed 14 Apr 2021
23. Download VMWare Workstation Player. VMWare. https://www.vmware.com/ca/products/workstation-player/workstation-player-evaluation.html. Accessed 14 Apr 2021
24. Project Gutenberg File. https://www.gutenberg.org/files/1342/1342-0.txt. Accessed 14 Apr 2021
25. Learning Container. https://www.learningcontainer.com/mp4-sample-video-files-download/#. Accessed 14 Apr 2021

26. SmallPdf. https://smallpdf.com/edit-pdf. Accessed 14 Apr 2021
27. FileSamples. https://filesamples.com/formats/mp3. Accessed 14 Apr 2021
28. FileSamples. https://filesamples.com/formats/mkv. Accessed 14 Apr 2021
29. Ibarra, P.: Restoring Microsoft Windows Deduplication Volumnes. Jungle Disk. https://www.jungledisk.com/blog/2018/06/18/restoring-windows-deduplication-enabled-volumes/. Accessed 14 Apr 2021

Author Index

Printed in the United States
by Baker & Taylor Publisher Services